GENETIC DIVERSITY AND NATURAL SELECTION

UNIVERSITY REVIEWS IN BIOLOGY

ALREADY PUBLISHED

1. The Behaviour of Arthropods — J. D. Carthy
2. The Biology of Hemichordata and Protochordata (*Hardback only*) — E. J. W. Barrington
3. The Physiology of Nematodes (*Hardback only*) — D. L. Lee
4. The Metabolism of Insects — Darcy Gilmour
5. Reproduction in the Insects — K. J. Davey
6. Cybernetics and Biology — F. H. George
7. The Physiology of Trematodes (*Hardback only*) — J. D. Smyth
8. Aspects of the Physiology of Crustacea — A. P. M. Lockwood
9. The Physiology of Sense Organs — DeF. Mellon, Jr.
10. The Pattern of Vertebrate Evolution — L. B. Halstead
11. The Physiology of Cestodes — J. D. Smyth
12. Insect Hormones — V. B. Wigglesworth
13. The Ovarian Cycle of Mammals — J. S. Perry
14. Comparative Physiology of Renal Excretion — J. Riegel

OTHER VOLUMES ARE IN PREPARATION

GENETIC DIVERSITY AND NATURAL SELECTION

JAMES MURRAY

Department of Biology
University of Virginia

HAFNER PUBLISHING COMPANY
NEW YORK

1972

To my Father

In the U.S.A.
HAFNER PUBLISHING COMPANY, INC.
866 Third Avenue
New York, N.Y. 10022

OLIVER & BOYD
Tweeddale Court
Edinburgh EH1 1YL
A Division of the Longman Group Ltd.

ISBN 0 05 002457 4 Hardback
ISBN 0 05 002457 X Paperback
First published 1972

Printed in Great Britain by
BELL AND BAIN LTD., GLASGOW

Preface

The past five years have been exciting ones for students of population genetics and evolution. The radical change in the general appreciation of the significance of polymorphic variation has had a profound influence on our view of many problems, some old, some new. The new light has of course had a stimulating effect on the flow of research, so much so that it becomes difficult to pause for reflection.

Nevertheless it seems that the new directions are sufficiently well defined to make worthwhile an attempt at synthesis. What I have tried to do in this book is to pull together those lines of research specifically relating to the action of natural selection on genetic variation. In doing so I have kept in mind my advanced undergraduate and beginning graduate students. I hope that this volume will provide them with an introduction to the literature of ecological genetics and evolution.

Many people have contributed to the making of this book and deserve my thanks. First and foremost, Elizabeth Murray has been involved at every stage with research, organization, editing, and support. Professor Bryan Clarke and Dr. D. A. West read an earlier draft and offered valuable suggestions. Mrs. Elizabeth Kater and Mrs. Lucy Parks have helped with the preparation of the manuscript. Mr. Mark Probst prepared the illustrations, and Mrs. Betty Kirkhart typed the manuscript.

For permission to use their published materials in this book, I should like to thank Professor W. F. Blair, Professor A. D. Bradshaw, Professor A. J. Cain, Professor B. C. Clarke, Dr. Verne Grant, Dr. Motoo Kimura, Dr. K. F. Koopman, Professor M. Lamotte, Dr. R. C. Lewontin, Dr. M. J. Littlejohn, Dr. S. J. O'Brien, Dr. Peter O'Donald, Dr. E. Paterniani, Madame Claudine Petit, Professor David Pimentel, Professor Alan Robertson, Professor F. W. Robertson, Professor P. M. Sheppard, Dr. J. A. Sved, Professor J. M. Thoday, and Professor Bruce Wallace. Also I should like to thank the editors of the *American Journal of Human Genetics*, the *Bulletin de la Société Zoologique de France*, *Evolution*, *Genetics*, *Heredity*, *Nature*, and the *Quarterly Review of Biology*, and the National Academy of Sciences, the Royal Society, the Systematics Association, the Cambridge University Press, the Yale University Press, the University of Chicago Press, and W. W. Norton and Company.

JAMES MURRAY

Contents

1 : Introduction

'. . . unless profitable variations do occur, natural selection can do nothing.' Charles Darwin: *On the Origin of Species.*[38]

The essential pre-condition for the occurrence of evolution in any population is the existence of heritable variation. Within the limits of a single generation, natural selection or other agencies may alter the level of expression of a character in a population; but unless the change is reflected in the offspring of the selected individuals, the effect will be only transient. For this reason, questions about the nature and extent of the genetic variation in natural populations are absolutely fundamental to evolutionary theory.

The crucial importance of heritable variation to the principle of natural selection was perfectly clear to Charles Darwin. He was convinced, moreover, that a great store of genetic variability was latent in most organisms. In the *Origin* he writes that 'the number and diversity of inheritable deviations of structure, both those of slight and those of considerable physiological importance is endless.'[38] (p. 12). Darwin's confidence here rests on observation. He is speaking from his first-hand experience both with domestic animals such as the pigeon, *Columba livia*, and with geographical variation, so beautifully exemplified in the finches of the Galapagos Islands.

Despite his convictions on the ubiquity of genetic variation, Darwin was acutely disturbed by the problem of its origin and maintenance. He faced a special difficulty, not always appreciated today, generated by the inadequate knowledge of hereditary transmission at that time.

Darwin accepted the contemporary idea of blending inheritance, the theory that the contributions of male and female parents are inextricably mingled in the offspring to produce individuals with intermediate character traits. No better explanation was currently available to account for the resemblance of offspring to both parents. Darwin was fully aware that such a blending of characteristics must inevitably result in the very rapid erosion of the variation within populations. He also accepted the consequent conclusion that the continued existence of such variation must necessarily entail its continual renewal in each generation.

But, as Fisher[62] has pointed out, Darwin was never wholly satisfied with this theory of inheritance. He realized that the reappearance of characters, which were absent in the parents but present in remote ancestors, can hardly be explained by blending inheritance. Darwin's hypothesis of 'a tendency in the young of each successive generation to produce the long-lost character . . . (which) . . . from unknown causes, sometimes prevails'[38] (p. 166) is clearly *ad hoc* and unsatisfactory. Domestic animals and plants presented an additional difficulty. If, as Darwin supposed, the changed and improved conditions of domestication were the cause of variability in domestic animals, then the effect should be most pronounced soon after domestication and should subside as blending inheritance takes its toll under the new stable conditions. On the contrary, Darwin realized that the oldest of the domesticated animals and plants, such as wheat, tend to be the most variable.

The special difficulties of blending inheritance were, of course, swept away by the discovery of Mendel's laws. Since 'hybrids' or heterozygotes represent only a temporary union of the genetic elements, the reappearance of ancestral characters in a regular manner is predicted by the Law of Segregation. There is thus no tendency inherent in the genetic system itself towards loss of genetic variation. This implicit property of Mendelian genetics is expressed explicitly as the Hardy–Weinberg Equilibrium.

The effect of the Mendelian revolution has been to bring about a curious reversal in the evolutionary roles of heredity and selection. Faced with the tremendous loss of variability by blending inheritance, Darwin tended to regard natural selection as an agent for delaying the loss and for preserving different variations in time and space. Darwinism is therefore a system in which genetics promotes uniformity and selection ensures diversity, while Neo-Darwinism is one in which the preserver of diversity is genetics and the agent promoting uniformity is selection.

The principal thesis of this book is that perhaps Darwin's position, although based on faulty genetic premises, may not have been so wide of the mark as many Neo-Darwinians have suggested. It is becoming increasingly evident that the idea of the 'wild type' and its mutants, a concept encouraged by the outstanding success of genetic studies with *Drosophila melanogaster*, may be seriously in error. Indeed, it is one of the curious ironies of the history of science that Mendelian genetics, while providing evolutionary theory with one of the keys to the maintenance of variation in populations, should nevertheless have led at least some geneticists to question the extent and importance of naturally occurring variation.

If, then, we are to achieve anything of a post-Neo-Darwinian synthesis of evolutionary theory, it is essential to discuss the new–old role of natural selection as a promoter of diversity in natural populations. It is the purpose of this book to attempt an assessment of the present state of knowledge of variation in populations and to inquire whether the concept of natural selection may not be of crucial importance in our understanding of the situation. Chapter 2 examines the data on polymorphism and attempts to reconcile them with conventional concepts of population genetics. Chapter 3 suggests that frequency-dependent selection may offer an alternative or additional answer to the puzzling problem of ubiquitous polymorphism. Chapters 4 through 6 discuss some of the ways in which this approach may throw new light on old evolutionary problems such as the nature of dominance, the origin of isolating mechanisms, and speciation.

Although these topics appear on the surface to be quite diverse, I find a common thread running through all of them. Each is concerned, on one level or another, with diversity in biological systems and with the interaction between genetic diversity and natural selection.

2 : Polymorphism

> 'And if there be any variability under nature, it would be an unaccountable fact if natural selection had not come into play. It has often been asserted, but the assertion is quite incapable of proof, that the amount of variation under nature is a strictly limited quantity.' Charles Darwin: *On the Origin of Species*.[38]

Variation is a characteristic and fundamental property of biological systems. Every level of organization displays variation in some parameters, in space or time, within or between cells, tissues, organisms, populations, and communities. The basic problem of the biologist is to explicate the nature, extent, and causes of this almost overwhelming complexity.

Usually the investigation of biological systems proceeds in three successive stages. The initial questions deal with the qualitative pattern of the variation. Next come inquiries about the extent of the variation, resulting in quantitative estimates. Finally there are questions about causative mechanisms. What are the forces that generate, conserve, restrict or destroy the variation within the system? The three levels of inquiry are by no means mutually exclusive, of course, but go along with one another in an irregularly overlapping sequence.

The study of population genetics is unusual in that the second and third aspects of the study of populations have largely been inverted. Because of the early application of mathematical analysis, so powerfully developed by Haldane, Fisher, and Wright, and because of the influence of the Darwinian theory of natural selection, the causal analysis of the forces affecting genes in populations has largely outrun the quantitative estimation of genetic variation. The last ten years have seen a re-awakening of interest in this problem of the amount of variation in natural populations. The emerging answers, derived first from indirect methods and more recently from direct estimates made possible by new techniques, have threatened to shake the foundations of population genetics. It will be the task of the next few years to reconcile the emerging pattern of variation with the structure of classical theory.

Indirect Estimates of Heterozygosity

The value of genetic polymorphism as a tool for the study of evolution in natural populations has long been recognised.[68] As a consequence, a very extensive literature has been built up cataloguing a great many polymorphic systems in a wide range of organisms (for an introduction, see [90]). Most of these earlier studies, however, deal with only one or a few polymorphisms in any species; and they are primarily concerned with the evolutionary dynamics of the limited system. They leave unanswered the question of whether the observed polymorphisms are special cases involving only a small proportion of the total number of genetic loci, or whether polymorphism is the normal condition at many loci.

In the absence of direct information two opposing views have been widely held. Arguing principally from the mathematical considerations of genetic load, Crow and his co-workers pictured the usual population as consisting largely of homozygotes with genetic variability maintained by recurrent mutation. Polymorphism, due to heterozygous advantage, would then be limited to a very few loci.[34] Muller[137] also expressed the view that such polymorphisms are temporary expediencies which simply tide the population over until a homozygote of equal or superior fitness appears.

The alternative view, upheld by Dobzhansky and his colleagues, derives from their extensive studies on the genetic architecture of *Drosophila*. It is 'that the adaptive 'norm' of a species established by natural selection is not genetically monolithic; it encompasses a large diversity of genetic elements, many of which show unexpected and unwelcome properties when homozygous.'[47]

The first concerted attempt to distinguish between what may be called the 'homozygous model' and the 'heterozygous model' of genetic structure was undertaken by Wallace[193] using the effects of randomly induced mutations on the viability of homozygotes in *Drosophila melanogaster*. It is a commonly observed phenomenon in *Drosophila* that flies which are made homozygous for chromosomes from wild populations show reduced viability. The reason for the reduction, however, is not immediately apparent. According to the homozygous model, the lowered fitness is the result of the unmasking of deleterious recessives. While this effect would also be found with the heterozygous model, there would be the additional disability of homozygosity at loci which show optimal fitness as heterozygotes.

Wallace argued that it might be possible to distinguish between the models on the basis of the response of flies, made homozygous for a particular second chromosome, to randomly induced mutations. Under

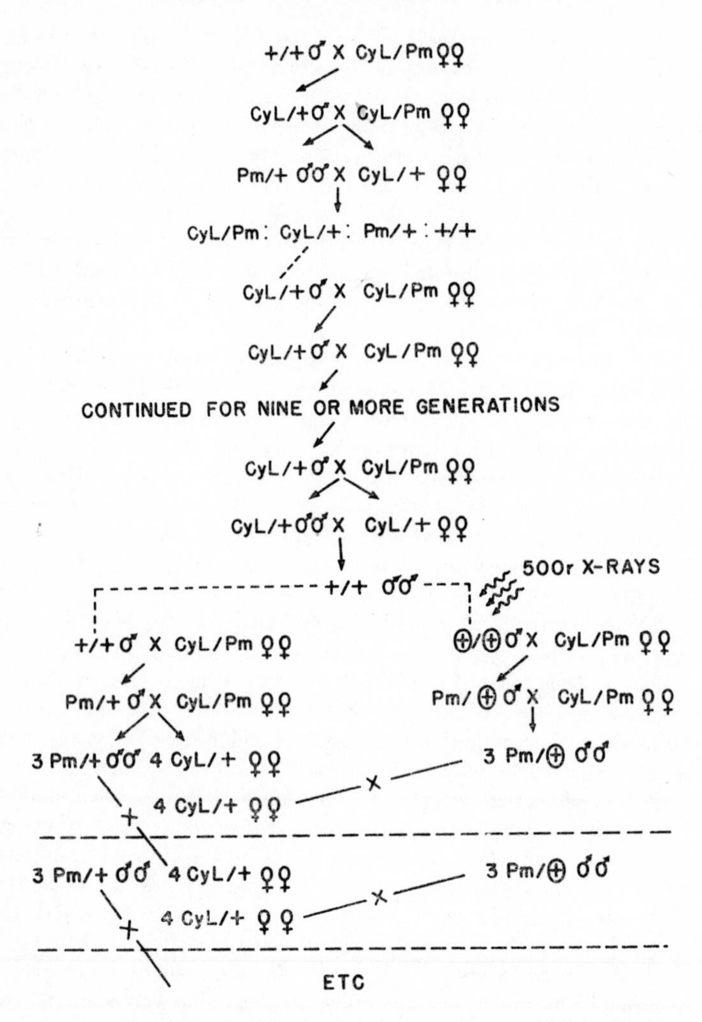

FIG. 1. A mating system devised to study the effects of mutations in heterozygous condition on the viability of flies made homozygous for quasi-normal second chromosomes. (After Wallace[193].)

the homozygous model, these second chromosomes would approximate the condition in natural populations, except for the homozygosity of a small proportion of deleterious alleles. Since deleterious alleles represent only a small fraction of the total, the probability that a mutation would increase viability is very small. The expected result of random mutation would therefore be to alter some of the optimal alleles to sub-optimal ones. The net effect of mutation should be to reduce still further the viability of the flies.

On the contrary, the heterozygous model postulates a number of possible optimal alleles at each locus producing maximal viability in the heterozygous condition. Hence when a second chromosome is made homozygous, the probability of a mutation to an allele with favourable interactions is high. It follows that in at least a proportion of cases, mutation should increase viability. The frequency with which increases and decreases occur should permit the assessment of the proportion of mutations which are favourable or unfavourable.

To test these predictions, Wallace extracted four quasinormal chromosomes from a laboratory population of flies which had a history of radiation but which had been maintained without radiation for almost two years prior to the experiment. These chromosomes were transferred into the genetic background of a Curly Lobe/Plum tester stock by continued backcrossing. Single Cy L/+ males were used in each generation to ensure that each experimental chromosome retained its individual identity. After at least nine generations of backcrossing, Cy L/+males and females were mated. Some of the +/+ sons were irradiated and some were used as controls. Finally Pm/+ males bearing an irradiated chromosome were crossed with Cy L/+ control females to produce four classes of offspring: Cy L/Pm, Pm/+, Cy L/+ irradiated, and +/+ irradiated. These were then compared with control crosses, similar in every respect except for the irradiation. The mating scheme is diagrammed in Fig. 1.

TABLE 1

The average relative viabilities of flies of different genotypes compared to CyL/Pm *as* 1·000. *Only those lines marked * actually carry an irradiated chromosome.* n *is the number of cultures on which the averages are based.* P *indicates the probability that there is no difference between the pairs of control and irradiated lines.* (*After Wallace*[193].)

	CyL/+	Pm/+	+/+	n
Irradiated	1·115*	1·137	1·033*	764
Control	1·094	1·146	1·008	766
P	0·04	0·15	0·02	

The results, shown in table 1, are very striking. As expected, the Pm/+ flies from the irradiated lines and the control lines do not show any differences, since the wild type chromosomes in the two lines are identical and unirradiated. The +/+ flies of the irradiated line, however, show markedly higher viability than those of the control line (P = 0·02). This is, of course, strong evidence in favour of the heterozygous model, although the magnitude of the effect is so large as to cause some concern. The problem is compounded by the fact that the Cy L/+ flies of the irradiated lines also show increased viability over the control Cy L/+ lines (P = 0·04). This result is unexpected under either model since both lines are already heterozygous, and hence the irradiation is unlikely to have increased the overall heterozygosity of that line.

Because of the unusual nature of the results, Wallace carried out an exhaustive analysis of the data, showing that although there was significant variation among the different chromosomes, experiments, and technicians caring for the flies, there was nevertheless no systematic bias detectable that might prejudice the data in the direction of higher fitness of the irradiated lines. Moreover in a subsequent experiment[194] the difference in the Cy L/+ lines disappeared. Consequently one can conclude that the experiment is inconsistent with the homozygous model and tends to support the heterozygous model. Considering that the induction of new mutations, and consequently heterozygosity, resulted in improved viability in over half the observed cases, Wallace suggested that individual wild *Drosophila* might be expected to be heterozygous at approximately 50% or more of all its loci.

Direct Estimates of Heterozygosity and Polymorphism

The great difficulty in the estimation of the amount of polymorphism in natural populations is the problem of identifying traits produced by specific loci for which there is no detectable genetic variation. With the techniques of classical genetic analysis, the observation of variation is the initial step in any study. Invariant loci simply remain undetected. In order to provide an unambiguous answer to questions about the proportion of loci which are polymorphic, some method of random sampling of the genome is required. Hubby and Lewontin[87] have suggested criteria for such a process: (1) Phenotypic differences caused by allelic substitution at single loci must be detectable in single individuals. (2) Allelic substitutions at one locus must be distinguishable from substitutions at other loci. (3) A substantial proportion of (ideally, all) allelic substitutions must be distinguishable from each other. (4)

Loci studied must be an unbiased sample of the genome with respect to physiological effects and degree of variation.

With the advent of the technique of electrophoresis and the development of extremely sensitive enzyme assays, these criteria may be met, at least approximately. The first two simply demand that one be able to identify clear-cut Mendelian differences. Where electrophoretic variants are found, they almost invariably display such a simple Mendelian inheritance. The principal assumption involved here is the extension of this conclusion to the equation of invariant electrophoretic bands with individual invariant loci. The third criterion is only partially met, since the detection of allelic substitutions is limited to those which affect the net charge of the protein under examination. Since neutral substitutions will be overlooked, the estimate of the proportion of variable loci will be a minimal value.

Finally, in any particular study, it is perhaps most difficult to be sure that the fourth criterion is being met. The technique of electrophoresis focuses attention either on enzymes, detectable by their activity, or on proteins present in sufficient quantity to be identified with general protein stains. At present there is no method of determining whether these two classes of proteins are a representative sample of the products of all the loci of the genome.

The first attempts to undertake a random sampling of the genome of an organism were undertaken by Hubby and Lewontin[87,114] with *Drosophila pseudoobscura* and by Harris[82] with man. In the *Drosophila* experiments, Hubby and Lewontin investigated a group of enzymes, each chosen solely on the basis of the existence of a good assay for the enzyme. In order to avoid any possible bias inherent in the choice of enzymes as a group, they added to their sample a number of larval proteins, the function of which was not determined.

The assays were carried out on strains of flies from six widely separated areas. Each culture was established from a single wild-caught inseminated female. Most of the strains had been maintained in mass culture in the laboratory for about five years at the time the experiments were carried out. These were supplemented by eleven additional strains from Strawberry Canyon, California, which had been in captivity for only two or three generations. Between 16 and 44 lines were assayed for each of the enzymes or other proteins studied. Of 21 loci which were sufficiently constant in their electrophoretic properties to be reasonably well characterized, Hubby and Lewontin detected genetic variants at nine loci. The variable loci included those determining six of ten enzymes assayed and three of eleven larval proteins.

For a quantitative study of the frequency of alleles in the populations represented by the various strains, Lewontin and Hubby[114] evidently found it necessary to eliminate some of these loci. They report data on eighteen loci, with six of eight enzymes and three of ten larval proteins variable. On the basis of these figures, and excluding two loci that were variable in only one population, they estimate that $\frac{7}{18}$, or 39%, of all loci in the genome are polymorphic over the whole species, that each population is polymorphic for an average of 30% of all loci, and that the average individual is heterozygous for perhaps 12% of its genome.

It must be emphasized that these estimates have been deliberately minimized. First, the method will not detect all of the differences in the structure of proteins. Second, the strains from any locality represent a very small sample of the gene pool of that population. Third, the long history of laboratory culture for most strains makes it unlikely that they retain all of the initial variability of the limited original sample. And finally, the estimate excludes the two loci which show variants in only one population.

There are two additional points of interest that arise from this study. It is remarkable that even among the strains established from single inseminated females, some loci are still segregating after a lapse of as much as seven years. One of the loci for the larval proteins (pt-8), for example, is segregating in at least one strain from each of the five major localities. The other point is that there seems to be no tendency for local fixation of different alleles, as one might expect if the alleles were functionally equivalent. This observation is of considerable importance in the interpretation of the polymorphism, as we shall see later.

The contemporaneous and independent studies by Harris[82] on human enzymes, chiefly those of erythrocytes, show a remarkably similar picture. The initial report indicated that two or three out of ten loci showed polymorphic variation. The study has now been extended[83] to eighteen enzymes. Six of these are polymorphic in the English populations examined. One of them, phosphoglucomutase, is controlled by three separate loci, so that overall 30 to 40% of the detectable loci are polymorphic. Some of the variation is restricted to particular sub-groups, Europeans showing polymorphism for 6 of the 20 and Africans, 7 of the 20 loci.

One of the limitations of the *Drosophila* studies cited above is that the tests were carried out largely on strains with long histories of laboratory culture. More recent studies have resolved this problem, although the question of random sampling remains. O'Brien and MacIntyre,[140] for example, tested a number of strains of *Drosophila melanogaster* and *D. simulans* recently taken from a number of different wild populations in the southeastern United States. One standard laboratory strain was included as a control. Only enzymes were assayed in this study, the inclusion of an enzyme depending entirely on the clarity and repeatability of the assay. Thus some enzymes previously known to be polymorphic were included.

The strains from any one locality represented a rather small sample of the gene pool of the population. If the founder females were singly inseminated, then the number of individual genes at each locus varied from 4 to 48 (3 to 36 for sex-linked loci). Moreover the assays were terminated as soon as polymorphism at a locus had been established. Therefore the *number* of alleles at a locus was almost certainly under-estimated.

Nevertheless, in *D. melanogaster*, polymorphism was detected at almost every locus investigated. Of ten enzymes assayed, only one, an acid phosphatase, was invariant in all populations. Interestingly enough, *D. simulans* was invariant for a different form. One other enzyme, a malic dehydrogenase, was polymorphic in a single population only. The other eight were found to be polymorphic in several populations. Table 2 summarizes the data. It may be seen that the average proportion of polymorphic loci in any population is 54% with the average individual heterozygous for approximately 23% of its genome. Significantly, the laboratory population shows very nearly the same degree of polymorphism.

TABLE 2

The proportion of loci polymorphic and the average proportion of heterozygosity per individual in populations of Drosophila melanogaster *and* D. simulans. (*After O'Brien and MacIntyre*[140].)

Species	Population	No. of Loci Polymorphic	Proportion of Loci Polymorphic	Proportion of Genome Heterozygous/ Individual
D. melanogaster	Ceres	3	0·30	0·134
	Painesville	5	0·50	0·212
	Mt. Sterling	8	0·80	0·339
	Mammoth Cave	7	0·70	0·271
	Red Top Mt.	6	0·60	0·269
	Columbia	4	0·40	0·163
	Manning	4	0·40	0·166
	Oxford	6	0·60	0·258
	Average (excluding Kaduna)	5·4	0·54	0·227
	Standard Kaduna	5	0·50	0·230
D. simulans	Columbia	2	0·20	0·070
	Manning	0	0·00	0·000

Although the data for *D. simulans* were based on very limited samples, they suggest that wild populations of this species are considerably less variable than those of *D. melanogaster*. This indication is borne out by Berger's[208] analysis of samples from five localities and four inbred laboratory strains. *D. simulans* showed no variation at loci controlling six enzymes, five of which were polymorphic in *D. melanogaster*.

The extension of the studies of *Drosophila pseudoobscura* to newly-caught strains[156] has made relatively little difference to the earlier estimates of polymorphism in this species. A survey of populations from both central and marginal areas of its North American range showed an overall high level of polymorphism (42% of the loci sampled), with generally similar gene frequencies throughout. Only the disjunct population in Bogotá, Colombia, showed significant restriction of the variation. Even in this population, 25% of the loci remained polymorphic. Taken together with the results from the laboratory strains, these data indicate that genetic diversity is not easily reduced in *D. pseudoobscura*.

Evidence for the occurrence of extensive polymorphism at the protein level continues to accumulate for more and more species. Manwell and Baker[119] have shown, for example, that the snails of the genus *Cepaea*, which are famous for their extensive polymorphisms of shell colour and pattern,[183] are just as variable in their enzyme phenotypes. *C. nemoralis* is polymorphic at 17 of 27 loci (63%) and *C. hortensis*, at 14 of 27 (52%). The house mouse, *Mus musculus*, shows 17 variable loci out of 41 (41%).[165] Even the 'relic' horseshoe crab, *Limulus polyphemus*, has nine polymorphic loci out of 25 (36%), indicating that morphological constancy over great spans of geological time does not necessarily indicate genetic impoverishment.[166]

An Independent Estimate of Heterozygosity in Man

A quite different method of investigating the amount of polymorphism in human populations has been developed by Lewontin[113] using blood-group antigens. The theoretical possibility has always existed of estimating the proportion of polymorphic loci by comparing the number of loci with generally occurring antigens with those where only rare variants (family antigens) are known. The problem with this method is obvious: widely distributed antigenic variants will be much more easily detected than those of very limited occurrence. Hence there will be a built-in bias toward the overestimation of the proportion of polymorphic loci.

Lewontin has suggested a clever escape from this bias. Since the ubiquitous polymorphisms are likely to be discovered first, the bias in favour of polymorphism should be greatest at first. Then as more and more loci are detected, the proportion of rare variants among those discovered should increase. In time the proportions of known loci of both classes should approach the true values for the population.

If the polymorphic loci are plotted as a proportion of all known loci year by year, the resulting curve should asymptotically approach the true proportion. Fig. 2 shows that this appears to have happened for human blood groups. Although the apparent equilibrium is rather dependent on the fairly recent discovery of the *Au* and *Xg* polymorphisms, it nevertheless seems fairly clear that the final value must be in the vicinity of 33% of the loci. Using gene frequency values for the English population, this represents an average heterozygosity of 16% per individual.

These values, derived by a different method from an entirely different set of loci, are surprisingly similar to Harris's estimates for human data cited above, and are in good agreement with values for several other species. Evidently, in a wide variety of animals polymorphism is the rule rather than the exception.

The Maintenance of Polymorphism

The confirmation of the existence of such widespread polymorphism has presented a challenge to the structure of classical population genetics. It has long been argued (*e.g.*[97]) that the properties of the gene pool prohibit the existence of polymorphism at more than a small fraction of the total number of loci in a population. Hence we are faced with the problem of reconciling theory with the observational data.

Attempts to solve this dilemma of variation have usually invoked one of two possibilities, either the selective neutrality of most of the alleles, and therefore also a high mutation rate, or else heterosis without the consequences predicted by the classical model. A third possibility, frequency-dependent selection, has been mentioned but hardly explored. The remainder of this chapter will be devoted to the first two suggestions, while the third will be dealt with at greater length in the following chapter.

Neutrality of Alleles

There are three quite separate problems which must be distinguished when considering the possibility that the observed variation is the result of allelic neutrality. One is the classical controversy over the mere existence of 'non-adaptive characters'. The second is the problem of the number of neutral alleles that can be maintained in a finite population. The third is, finally, the more limited question of whether

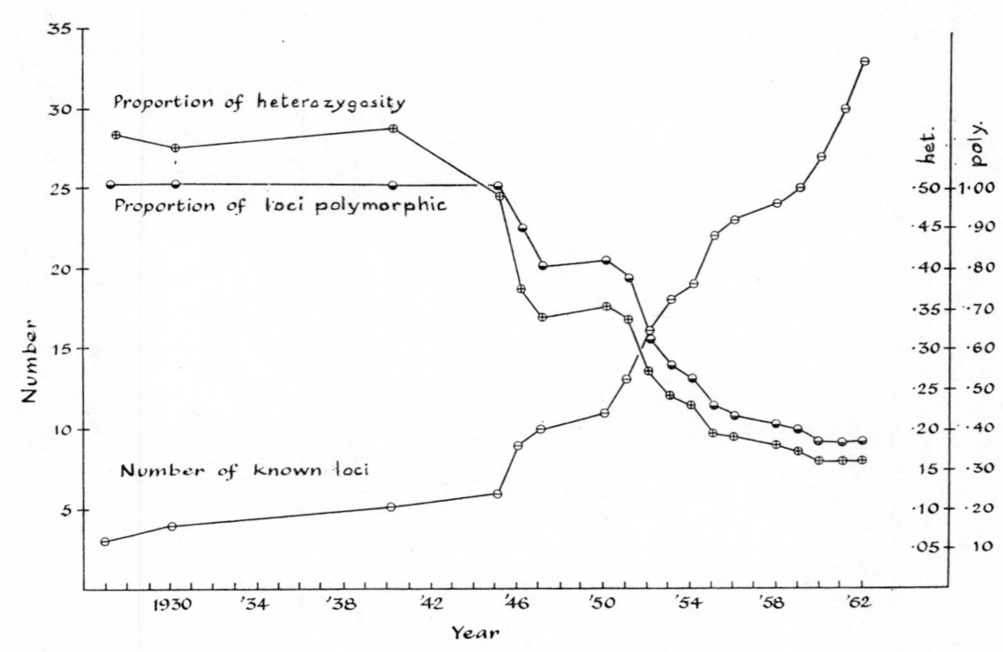

FIG. 2. Information on human blood group antigens expressed cumulatively year by year, showing the number of known loci, the proportion of loci that are polymorphic, and the proportion of heterozygosity based on gene frequencies of the English population. (After Lewontin[113].)

the particular alleles detected by electrophoretic studies of natural populations are selectively neutral.

The general problem of neutral characters, first cast into the terms of population genetics by Fisher[62] and Wright,[204] has recently been given a new dimension by the working out of the genetic code. The occurrence of redundancy in the code, where a single amino acid may be specified by more than one DNA triplet codon, implies the possibility of allelic differences which are not expressed at all in functional proteins. The mutant phenotype is thus restricted to the DNA sequence itself. While the redundancy itself is almost certainly an adaptive character,[72] it seems likely that in most instances changes within the redundant groups are neutral. That this is not invariably the case, however, is indicated by the example of arginine in mammalian DNA. Although six-times redundant, this amino acid seems to be regularly specified by only two of the six codons.[181]

At the level of amino acid substitutions the evidence points to a spectrum of effects. Comparative studies of homologous proteins from different species indicate that some regions of each molecule are more susceptible to evolutionary change than others. For example, if one tabulates the number of different amino acids found in each position of the cytochrome c molecules of different species,[100] the distribution is markedly non-random ($P < 0{\cdot}001$), with more than twice as many invariant positions as are predicted by the Poisson distribution. In addition, the range of effects of substitutions at potentially variable loci is considerable. As Goldberg and Wittes[72] have pointed out, the grouping of triplets in the genetic code is such that mutations are much more likely to bring about the substitution of a physically and chemically similar amino acid than would be expected on the basis of random changes. Hence the adaptive properties of the genetic code itself ensure the neutrality, or near neutrality, of as great a proportion of mutational changes as possible.

Provided that at least some neutral mutations occur, what will be the probable fate of the alleles which are so generated? Kimura and Crow[96,97] have discussed in detail the limits of variation that may be sustained in finite populations. They consider a population of N diploid individuals with an average mutation rate, u, per locus. The possibilities of neutral mutation are considered to be so large that each new mutant is treated as unique, so that $2Nu$ new alleles are generated at each locus in every generation. In time an equilibrium will be established between the gain of new alleles by mutation and their loss through genetic drift. The equilibrium number of alleles can be estimated by considering the inbreeding coefficient, F, which is defined as the probability that two uniting alleles are identical by descent from the same allele in a previous generation. An approximate equilibrium value for the inbreeding coefficient is

$$F \simeq \frac{1}{4N_e u + 1}$$

where N_e is the effective size of the population.

In the model developed by Kimura and Crow, F will represent the proportion of homozygotes in the population. If all alleles are equally frequent, then F will also be the reciprocal of the number of alleles. This number, n_e, is defined as the *effective number* of alleles in the population.

$$n_e = \frac{1}{F} \simeq 4N_e u + 1$$

The average proportion of heterozygosity and the effective number of alleles for varying population sizes and mutation rates are shown in table 3. It may be seen that to maintain much more than 4% heterozygosity and much more than one allele at a locus, the effective population

TABLE 3

The average proportion of homozygosity (upper figure) and the effective number of alleles (lower figure) in a randomly mating population of effective size N_e. *The alleles are assumed to be neutral and each mutation unique. (After Kimura and Crow*[97].)

	Effective population number, N_e					
Mutation rate, u	10^2	10^3	10^4	10^5	10^6	10^7
10^{-4}	0·96	0·71	0·20	0·024	0·0025	0·00025
	1·04	1·4	5·0	41	401	4001
10^{-5}	0·996	0·96	0·71	0·20	0·024	0·0025
	1·004	1·04	1·4	5·0	41	401
10^{-6}	0·9996	0·996	0·96	0·71	0·20	0·024
	1·0004	1·004	1·04	1·4	5·0	41
10^{-7}	0·99996	0·9996	0·996	0·96	0·71	0·20
	1·00004	1·0004	1·004	1·04	1·4	5·0

sizes must be rather large (10^4–10^5) or else the mutation rates quite high (10^{-4}). It must also be emphasized that this model maximizes the estimate of variation. If the number of possible alleles is restricted, then the proportion of heterozygotes may be very much reduced. It seems, therefore, that the observed variation in natural populations places a severe strain on the ability of the model to explain the data.

The final question to be considered is the direct one: whether the particular alleles identified by the electrophoretic method can belong to a potential class of neutral alleles. The evidence is by no means complete or even extensive; but such as it is, it suggests that the alleles at these loci are subject to selection. The presumption is strong even in the initial experiments of Lewontin and Hubby[114] on *Drosophila pseudoobscura.* There are at least three observations which are difficult to accommodate within a hypothesis of neutral alleles.

In the first place the observed pattern of variation is not that which is expected if the variation is generated by mutation and genetic drift. There are loci which are monomorphic in all populations, there are loci with rare variants, and there are loci which are so variable that no 'wild type' can be identified. However, there is a kind of differentiation that is conspicuously absent, *i.e.*, a pattern of variation with the fixation of different alleles in different localities. This last pattern is the expected outcome of allelic neutrality.

A second significant point is that the degree of polymorphism does not seem to be very sensitive to changes in population size. The effects of genetic drift should, of course, be more apparent in populations of restricted size. Among the populations sampled by Lewontin and Hubby, those from Wildrose and Cimarron probably have effective population sizes between one-fifth and one-tenth the size of populations from optimal areas such as Mt. San Jacinto, where Dobzhansky and Wright[48] estimated the effective size as 500 to 1000 individuals. There is no indication, however, of any restriction on the variation in the sparsely occupied areas. In fact they are the only populations showing variation in alkaline phosphatase-6 and -7.

Finally there is the observation that many of the single strains assayed by Lewontin and Hubby are segregating after years in the laboratory. This is quite remarkable in view of the extreme 'founder effect' expected in the establishment of each strain from a single inseminated female.

More direct, if still rather limited, information on the status of the electrophoretic variants comes from a study of their distribution in the chromosome inversions of *Drosophila pseudoobscura*, *D. persimilis*, and *D. miranda*. These species show great variety in the gene arrangements of their third chromosomes. It has proven possible, by considering overlapping inversions, to reconstruct the order in which the inversions have arisen one from the other, although the *direction* of change cannot be established with certainty (for a pictorial review, see[195]). Fig. 3 illustrates the relationships between some of the common gene arrangements of the three species. It may be seen that the Standard (ST) arrangement is common to both *D. pseudoobscura* and *persimilis* and that Pikes Peak (PP) and Arrowhead (AR) in *pseudoobscura* are each related to ST by a one-step transition. On the other hand, Santa Cruz (SC) is removed from ST by two steps involving an as yet un-

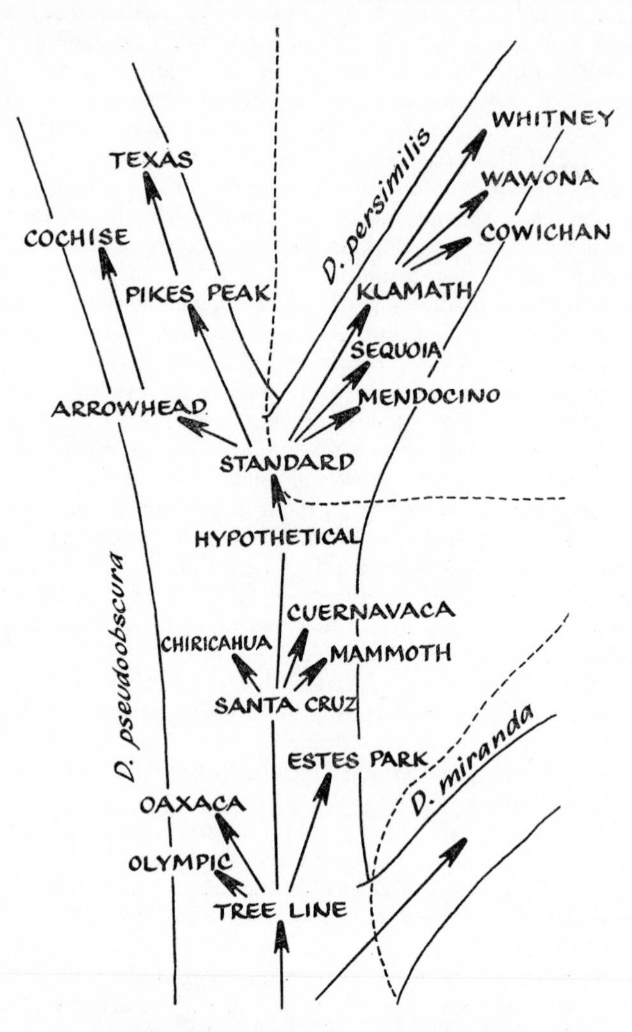

FIG. 3. Relationships between some of the common gene arrangements on third chromosomes of *Drosophila pseudoobscura*, *D. persimilis*, and *D. miranda*. The sequence of changes, but not the direction, can be determined from the study of overlapping inversions. (After Wallace[195].)

discovered hypothetical arrangement. SC is the centre of a cluster, including Chiricahua (CH) and Tree Line (TL), both one step from SC.

Prakash and Lewontin[155] have studied the frequencies of alleles at two loci on chromosome III and have shown that the distribution of frequencies is by no means random. Table 4 shows not only that

TABLE 4

The frequencies of alleles at locus pt-10 in different gene arrangements of Drosophila pseudoobscura *from different localities. (After Prakash and Lewontin*[155]*.)*

Gene arrangement	Allele	Population				
		Mather	Strawberry Canyon	Mesa Verde	Austin	Bogotá
		n = 20	*n* = 33	*n* = 1	*n* = 5	
Standard	1·04	1·00	1·00	X	1·00	—
	1·06	—	—	—	—	—
		n = 20	*n* = 6	*n* = 240	*n* = 7	
Arrowhead	0·94	0·10	—	—	—	—
	1·02	—	—	0·02	—	—
	1·04	0·90	1·00	0·97	1·00	—
	1·06	—	—	0·01	—	—
		n = 26			*n* = 69	
Pikes Peak	1·02	—	—	—	0·015	—
	1·04	1·00	—	1·00	0·985	—
			n = 2			*n* = 38
Santa Cruz	1·06	—	1·00	—	—	1·00
		n = 14	*n* = 11		*n* = 1	
Chiricahua	1·04	0·50	—	—	—	—
	1·06	0·50	1·00	—	X	—
		n = 20	*n* = 20		*n* = 3	*n* = 38
Treeline	1·04	—	—	—	0·33	—
	1·06	1·00	1·00	—	0·66	1·00
					n = 1	
Olympic	1·06				X	

X indicates presence of allele where frequency is not estimated; *n* = number of chromosomes.

particular alleles are characteristic of particular inversions over wide geographic areas but also that related clusters of inversions tend to share the same characteristic allele. These relationships transcend the limits of the species. At the pt-10 locus, for example, one allele is almost fixed in the ST cluster and also in *D. persimilis*, while another is at high frequency in the SC cluster. While the results with the other locus, α-amylase, are not as clear-cut as with pt-10, the general picture is much the same. In this case *D. persimilis* has four alleles, one of which is the predominant allele of the ST group.

The relationships of the alleles to the inversion types, extending as they do across species lines, imply great stability of the characteristic frequencies. Moreover, these cannot be simply the frozen relics of former random differentiation, since alternate alleles exist at low frequencies in most inversion types. Hence the conclusion is inescapable that at least in these instances, the alleles are not neutral but are held by natural selection at optimal frequencies for the particular constellation of genes of which they are a part.

Heterosis

If neutrality is rejected as a probable explanation for the observed variation in natural populations, the problem then becomes the determination of the mode of action of selection in maintaining the polymorphisms. Following historical precedent it is natural to think first of a selective advantage for the heterozygotes, or heterosis. However, as Lewontin and Hubby[114] have pointed out there are two different and rather serious objections to heterosis as a general mechanism. On the one hand it is difficult to imagine the physiological basis for so many different heterotic effects; and on the other, the total amount of selection required to maintain thousands of loci in a heterozygous state may be beyond the reproductive capacity of the average species.

To illustrate the dilemma, Lewontin and Hubby propose a model *Drosophila* population with 2000 loci each with two alleles. The homozygote for each allele at the first locus has a fitness of 98% of that of the heterozygote. This represents a segregational load or reduction in the mean fitness of the population of 1% for that locus. There will be an equivalent load for each of the 2000 loci. The reproductive potential of the population is therefore calculated to be $0{\cdot}99^{2000}$ or about 10^{-9}. It seems quite unreasonable that any *Drosophila* population could be realizing so little of its reproductive potential.

This model for the multiplicative action of selection at many loci has been challenged by a number of persons on rather similar grounds.[99,131,182] Basically the argument is that it is unrealistic to expect that selection operates via the direct multiplicative effects of the genes, as the model proposes. Rather, the fitness of any particular individual is only a function of the multiplicative probability, the probability of survival depending also on the selective values of the other members of the population. The standard against which each individual will compete, is that of the majority of the population clustering around the mean number of heterozygous loci. Not only will individuals approaching maximal heterozygosity be extremely rare, but also there is likely to be a limit to the amount by which the fitness of these exceptional individuals may exceed the mean fitness. To turn the *Drosophila* argument around, the super-heterozygote could hardly produce even ten times as many offspring as the individual at the population mean.

Sved and his colleagues illustrate their model with the example of a population containing 10,000 segregating loci. Each locus has two alleles with the relative fitness of the genotypes as follows:

Genotypes	A_1A_1	A_1A_2	A_2A_2
Fitness	$1-s$	1	$1-t$

where $s = t = 0{\cdot}01$. The reduction in mean fitness of the population as a result of selection at this locus is $st/(s+t)$, or $0{\cdot}005$, so that the mean fitness is $0{\cdot}995$.

In this symmetrical example the proportion of individuals with any particular number of heterozygous loci will be given by the binomial distribution with a mean of 5000 loci and a variance $pqn = 2500$. Fig. 4 illustrates the normal approximation of this binomial distribution. The absolute selective value of any individual is found by taking the fitness of the homozygote for one locus and raising it to the power representing the number of loci for which the individual is homozygous. The individual with the mean number of homozygous loci thus has the fitness of $0{\cdot}99^{5000}$, an absurdly small value. However these values may be converted into *relative* selective values by dividing by the mean fitness of the population. The relative fitness of the individual at the mean then becomes $0{\cdot}99^{5000}/0{\cdot}995^{10,000}$, or $0{\cdot}881$.

Where then on the distribution would an arbitrary limit of 10 for the fitness be found? If an individual is heterozygous for 5242 loci, then the relative fitness becomes $0{\cdot}99^{4758}/0{\cdot}995^{10,000} \simeq 10$. This point is very nearly five standard deviations (242 loci) from the mean and would be surpassed by fewer than one individual in 10^6. Thus, although postulating a limit to the maximum fitness must necessarily reduce the selective advantage of the heterozygote at each locus, the reduction is so small as to be negligible.

A further problem which has been raised by Sved, Reed, and Bodmer[182] is the discrepancy between expected and observed values for inbreeding depression. If many loci are being maintained in a heterozygous state by heterosis, then a large decrease in fitness is to be expected on inbreeding. Estimates of inbreeding depression, however, show a rather moderate decrease. Nevertheless, if the model of relative advantages is retained, then there is no reason to expect undue inbreeding depression. As Milkman[131] has pointed out, the competitive model does not preclude the existence even of a population of complete homozygotes.

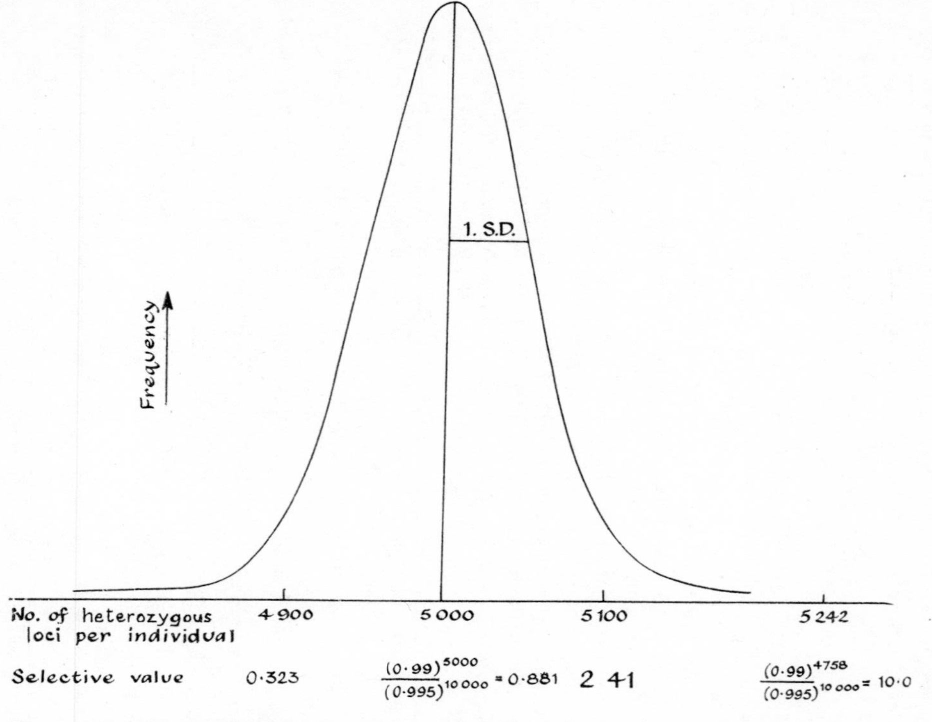

FIG. 4. The normal approximation of the binomial distribution of the number of heterozygous loci per individual in a population where 10 000 loci are polymorphic and where the coefficient of selection associated with each homozygous locus is 0·01. (After Sved, Reed, and Bodmer[182].)

In all these considerations, the distinction being drawn is that between what Wallace[196] has termed 'hard' and 'soft' selection. Hard selection may be said to result from direct interactions between the individual and its environment. For example, a petri plate containing streptomycin allows the growth of streptomycin-resistant organisms only, irrespective of the number of wild-type individuals present. On the other hand, soft selection is the result of competitive interactions between individuals of the population such that for a given unit of environment a fairly constant number of individuals will be produced. Wallace[196] has shown, for example, that the variance of the number of *Drosophila melanogaster* produced per vial of medium is significantly smaller than expected by chance. There is therefore a density-dependent control of the number of emerging flies. Applying this concept to the problem of heterosis, it appears that selection proceeds simply by culling that portion of the population with the largest number of homozygous loci.

Conclusions

In summary, therefore, the problem of the maintenance of large numbers of segregating loci by heterosis does not appear to be insuperable. In fact, the range of common fitness values appears to be limited, provided that an upper boundary is set to the maximum allowable fitness. The numbers of individuals which exceed such a limit is likely to be so small that they do not significantly reduce the heterosis per locus. The alternative possibility of neutrality of alleles would seem to have little to recommend it.

The remaining problem, which has hardly been considered so far, is the specification of the components of fitness affected by the heterotic selection. This task is enormous. It can only be met by a joint attack combining the resources of protein chemists and ecological geneticists, working together on the functional properties of enzymes and on the distribution of alleles in space and time.

3 : Frequency-dependent Selection

> 'Probably a very small biochemical change will give a host species a substantial degree of resistance to a highly adapted microorganism. This has an important evolutionary effect. It means that it is an advantage to the individual to possess a rare biochemical phenotype. For just because of its rarity it will be resistant to diseases which attack the majority of its fellows. And it means that it is an advantage to a species to be biochemically diverse, and even to be mutable as regards genes concerned in disease resistance.' J. B. S. Haldane: *Disease and Evolution.*[78]

Too often in the past the subject of genetic polymorphism has been discussed as if heterozygous advantage were the only means whereby a stable equilibrium can be maintained (*e.g.*[134]). Although, as we have seen, heterosis does offer a not unreasonable explanation of the ubiquitous polymorphism found in natural populations, a major alternative hypothesis exists in *frequency-dependent selection.* This mode of selection is not only capable of maintaining a balanced polymorphism, but it also offers the distinct advantage that under certain conditions segregational load may be eliminated. Moreover, frequency-dependent selection is a concept of very general application. Many special conditions that have been shown to lead to the stabilization of gene frequencies owe their particular properties to some inherent relationship between frequency and fitness.

Much of the structure of classical population genetics is built on the assumption that the fitness of an individual is a fixed property of that type in any given set of environmental conditions. In many cases this will be true only if the genetic structure of the population is considered to be a part of the total environment. For it is becoming increasingly apparent that fitness often varies according to the frequency of the different kinds of individuals within the population. In order to gain some measure of the importance of this mode of selection I shall first outline some of the theoretical properties of frequency-dependent relationships and then discuss some examples from natural populations.

The Theory of Frequency-dependent Selection

In his classic work *The Genetical Theory of Natural Selection*, Fisher[62] has outlined the general conditions under which a balanced polymorphism will be maintained. With the two alleles A_1 and A_2 at a single locus in a population at frequencies p and q respectively, the distribution of genotypes and their fitness may be indicated as:

Genotypes:	A_1A_1	A_1A_2	A_2A_2
Initial Frequency:	p^2	$2pq$	q^2
Fitness:	a	b	c
Frequency after Selection:	ap^2	$2bpq$	cq^2

After selection the frequency of allele A_2 becomes:

$$q_1 = \frac{bpq+cq^2}{ap^2+2bpq+cq^2}.$$

And the change of gene frequency is:

$$\Delta q = \frac{bpq+cq^2}{ap^2+2bpq+cq^2} - q$$

or

$$= \frac{pq[b-a-q(2b-a-c)]}{ap^2+2bpq+cq^2}.$$

It may be noted here that since both pq and also $ap^2+2bpq+cq^2$ must be positive, then the sign of Δq is determined by

$$[b-a-q(2b-a-c)]$$[112].

At equilibrium Δq becomes zero and the ratios of the gene frequencies are the same before and after selection such that:

$$\frac{p}{q} = \frac{ap^2+bpq}{bpq+cq^2}$$

or

$$ap+bq = bp+cq.$$

Substituting $1-q$ for p, this expression becomes:

$$\hat{q} = \frac{b-a}{2b-a-c}.$$

Thus the equilibrium gene frequency, $\hat{q}$, depends not on the initial value of q but only on the coefficients a, b and c.

The quantities a, b and c were denoted by Fisher[62] as frequency-independent selective values (although he recognized cases of frequency-dependent selection as will be seen below) such that stable equilibrium necessitated a value of b greater than a and c. More recently, however, Lewontin[112] has generalized this model simply by treating these entities as 'weights' of unspecified content. They may, therefore, encompass not only the effects of selection, both frequency-dependent and independent, but also mutation rates, migration rates, and the coefficient of inbreeding. These forces may be dealt with singly or in combination.

The stability of an equilibrium may be investigated by determining the rate of change of Δq with respect to q at or near the equilibrium point $\hat{q}$. There are six possibilities,[112] which are listed in table 5 together with the expected consequences.

TABLE 5

The rate of change of Δq with respect to q and the consequences to be expected in each case. (After Lewontin[112].)

$\frac{d\Delta q}{dq}$	Result
>0	No equilibrium; directional change leading to fixation.
0	Neutrality; no further change from the initial value of q.
0 to -1	Stable equilibrium; if disturbed, q returns to the equilibrium directly.
-1 to -2	Damped oscillation; q returns to the equilibrium point by overshooting at each oscillation.
-2	Uniform oscillation; no tendency for the oscillations to increase or decrease in magnitude.
<-2	Increasing oscillation; eventually leading to fixation.

In practice the stability of an equilibrium may be most easily determined by examining the sign of the derivative

$$\frac{d}{dq}[b-a-q(2b-a-c)].$$

This quantity will be positive if the equilibrium is unstable and negative if it is stable. The importance of this development is that it permits the demonstration, by means of simple numerical examples, that heterozygous advantage is neither a necessary nor sufficient condition for a balanced polymorphism.[112] Frequency-dependent selection may provide stability even though the heterozygote is selected against, or it may render unstable an equilibrium in which the heterozygote has the highest fitness.

Clarke and O'Donald[24] have extended the treatment of frequency-dependent selection to take into account interactions with frequency-independent selection and the effects of dominance. In all of their models the frequency-dependent component represents a constant relation between phenotype frequency and selective value. The models for frequency-dependent selection with and without dominance are quite straightforward. In the former, the heterozygote is at a disadvantage, but the equilibrium is stable with $\hat{q} = 0{\cdot}5$. In the latter, the dominant homozygote and heterozygote are confounded with the resulting equilibrium at $\hat{q} = \sqrt{0{\cdot}5}$. It is the interaction with frequency-independent components of selection which introduces interesting complications. With no dominance and some combinations of selective components of the two types there may be three equilibrium points, two of which are stable. At all three points the heterozygotes are at an overall selective disadvantage. Finally, with dominance added to this scheme, there may also be three equilibrium points, but only one is stable.[24]

The possibility of oscillation in a frequency-dependent system is an intriguing one. Haldane and Jayakar[79] have pointed out that here is a case, unusual in population genetics, in which there are large differences between a breeding system with discrete as opposed to overlapping generations. Oscillation by severe frequency-dependent selection will be minimized by the damping effect of overlapping generations. It would seem that the likelihood of oscillation is small; although with characteristic ingenuity Haldane and Jayakar suggest a hypothetical example in which a type that has the highest fitness in all other respects is subject to severe epidemic disease.

Cook[28] has also considered the effects of overall intensity of selection when both frequency-dependent and independent selection are operating. An important conclusion is that in such a mixed system the approach to equilibrium is likely to be delayed. Hence substantial frequency-dependent selection might be expected to lead to a wide range of gene frequencies in a series of polymorphic populations.

Frequency-dependent selection, then, if it is not unduly severe, is a force capable of maintaining a stable polymorphism in a population. Let us now examine some of the agencies which may act in this manner in natural populations. It will become apparent that many of the forces, often considered to be distinct mechanisms for the maintenance of polymorphism, may share the common denominator of frequency-dependent selection.

Non-random Mating

Perhaps the most familiar examples of polymorphisms maintained by frequency-dependent selection are cases of disassortative mating, or the tendency of unlike types to mate more frequently than expected by chance. Outbreeding mechanisms in general fall in this class, with sex as a common example.

The question of the influence of selection on the sex-ratio is one which completely mystified Charles Darwin.[39] Although he recognized a widespread tendency to produce a sex ratio of 1 (males/females), he was unable to discover an explanation in terms of individual selective advantage. He reasoned that since selection operates through the production of greater overall numbers of offspring, it should be immaterial whether these are males or females.

The key to the sex-ratio problem is a frequency-dependent selective mechanism first recognized by Fisher.[62] In a population with obligate cross fertilization, the parental contribution to the production of each sex in the next generation must necessarily be equal. Therefore the greatest return for invested biological 'capital' will be realized from the production of whichever sex is in the minority. Hence there will be a general tendency for the primary sex ratio to approximate unity, with the exact point of equilibrium adjusted by selection to compensate for greater losses in the less viable sex. Fisher's argument has recently been clarified and elaborated by Bodmer and Edwards[5] and by Williams[199] who emphasized that the sex ratio would be stabilized at 1 at the zygote stage except for the complication of parental care. The extra investment entailed by parental care results, for the less viable sex, in a higher cost per individual reared but a lower cost per individual born. Thus the equilibrium point for the sex ratio of 1 should be reached sometime between birth and the achievement of independence from the parents. Normally this would be only slightly before the end of the period of parental care.

The robust nature of the sex-ratio equilibrium has been emphasized by Shaw[167] who considered the possibility of disturbances introduced by polygamy and female infanticide. Neither of these conditions will be expected to affect the sex ratio, despite the argument put forward by Darwin in *The Descent of Man*.[39] Discussing Marshall's account of infanticide among the Todas of India, Darwin attempts to show that the death of female children produces selection for the production of males. Shaw shows that the sex ratio of the population will be unaffected, and that only in the case of unusually restricted population size coupled with incest prohibition will there be any effect on ratios within families.

Hamilton[80] has recently criticized Fisher's theory of sex ratio. He points out that there are cases in which selection will not automatically maintain the equality of the sexes. Notably there is the condition of meiotic drive in which one of the sex chromosomes is incorporated into functional gametes more often than the other. Either a driving X or Y, if unchecked, will bring about a change of sex ratio ending in extinction, the driving Y producing its effect much more rapidly. Driving X chromosomes have been described in wild populations of *Drosophila*,[70] and an analogous situation exists in the triploid female salamanders of the *Ambystoma jeffersonianum* complex where fertilization serves merely to initiate parthenogenetic development of more triploid females.[189] There is even a suggestion that some strains of *Aedes aegyptii*[85] harbour a suppressed driving Y. It is obvious that the evolutionary defence of a species against a driving sex chromosome lies in the suppression of the expression of meiotic drive. It is a corollary of Fisher's theory that the selection in favour of suppression would be intense. Indeed, Hamilton offers the suggestion that this effect may provide an explanation for the relative inertness of the Y chromosome, a genetically active Y representing a constant threat to the species.

Another mechanism whereby departures from a sex ratio of 1 may occur has been postulated in a number of diœcious plants. Correns[30] found that in *Melandrium album* (= *Silene alba*) sex ratio could be altered by controlling the amount of pollen reaching the pistillate flowers. Small amounts of pollen resulted in offspring with a sex ratio of 1, while large amounts produced an excess of pistillate plants. The effect seems to be produced by differential growth of the X- and Y-bearing pollen tubes. Darlington[37] concluded that this might be a mechanism whereby natural selection adjusts the sex ratio of the population to achieve maximum reproductive efficiency. In this case selection would be acting at the level of the population rather than on the individuals. Since Williams[199] has recently severely criticized the theory of group selection, it is of considerable interest that the case of *Melandrium album* has been investigated in the field.

The sex ratios of both wild and experimental populations of *Melandrium* have been studied by Mulcahy.[135] Although the results are not clear cut, and are further obscured by the concentration on flower ratios rather than sex ratios, several points stand out. First, in the three wild populations for which complete data are available, the sex ratios are consistently less than 1 (0·68, 0·60, 0·32). Second, in the single set of experimental populations tested for sex ratio of the offspring, there is a suggestion that populations with high sex ratios tend to produce excess pistillate flowers in the succeeding generations. Finally, the optimal sex ratio for maximum seed production seems to be much lower than the ratios in wild populations. It appears, therefore, that selection of individuals is preventing the population from achieving an optimal sex ratio, although ratios of less than 1 are usual. Further study of this situation should be rewarding.

A system which has the same effect as sex in enforcing disassortative mating and which therefore entails frequency-dependent selection is found in the self-sterility alleles of higher plants. In *Nicotiana*, for example, East and Mangelsdorf[49] explained the inability of hermaphroditic plants to fertilize themselves as the consequence of a series of alleles at a sterility locus S. Pollen bearing either of the alleles of the receptive plant is inhibited so that the pollen tubes grow very slowly if at all. Thus a plant of S_1S_2 genotype cannot be fertilized by either S_1 or S_2 pollen, but must receive S_3, S_4, or some other type. It follows that pollen bearing relatively rare alleles will be selectively favoured over pollen with common alleles.

The question of how many such alleles may be maintained in a population of limited size has been a recurring source of controversy ever since the discovery by Emerson[57,58] that the incompatibility system of *Oenothera organensis* comprises no less than 45 alleles. This is altogether remarkable since Emerson also estimated the total population of this relict species to be about 500 individuals. Wright's initial paper[205] on this subject has provoked a long, involved mathematical exchange with Fisher and Moran (see [206] for references). Wright has concluded that a population of 500, with a mutation rate of 5×10^{-7}, should contain only about 12 alleles.

More recently Crosby[33] has developed a new approach to the problem which avoids the mathematical intricacies of Wright's and Fisher's analyses. Using computer simulation techniques he has investigated the consequences of various combinations of assumptions, *i.e.* annual *vs.* perennial habit, high *vs.* moderate mutation rate, and varying degrees of isolation. The results, which are in broad agreement with Wright's model, indicate that the situation described by Emerson cannot possibly represent a steady state. The question of interest then is not how many alleles may eventually remain in the population but at what rate is the loss of alleles occurring and how long has the process been going on? The conclusion is inescapable that the situation in 1939 was the result of a disastrous reduction in the population of *Oenothera organensis*. From comparison with his models, Crosby suggests that this must have been a relatively recent event, probably within the last nine generations. With a long-lived perennial the time span could be of the order of 100 years, compatible with Emerson's suggestion that grazing may have been the agent responsible for the decline.

Examples of frequency-dependent selection produced by disassortative mating, apart from specific outbreeding mechanisms, are not very common. There are a few cases, however, where the persistence of a polymorphism may depend on this means. In the intensively studied moth *Panaxia dominula* (for references see [170]), a pair of alleles produces three visibly different phenotypes, the common homozygote, the heterozygote known as var. *medionigra*, and the rare homozygote, var. *bimacula*. The three forms differ in several respects, such as viability, fertility, and date of emergence, but perhaps the most striking aspect of the polymorphism is the tendency towards disassortative mating. In experiments to test mating preferences, Sheppard and Cook[170] have shown that both males and females of the *dominula* type prefer *medionigra* mates to *dominula*, although in the reverse combination *medionigra* shows no preference. In addition females of *bimacula* choose *medionigra* over *bimacula*, and females of *medionigra* prefer *bimacula*. Of 199 choice experiments undertaken, 126 resulted in unlike pairings and only 73 in like pairings ($X^2_{(1)} = 14{\cdot}12$, $P < 0{\cdot}001$). These experiments included material from more than one colony, indicating than the phenomenon is not a purely local one.

The effect of the mating preferences in *Panaxia* is enhanced by the fact that males are capable of mating many times while females mate only once. Hence there will be powerful frequency-dependent selection favouring *medionigra* so long as the gene producing it is rare. While it remains to be demonstrated that the effect of disassortative mating is great enough in natural populations to maintain the polymorphism, and although there are a number of other components of selection affecting this locus,[200] nevertheless the mating preference system is one which may be expected to contribute to the stability of the polymorphism.

In the case of *Panaxia*, there is no evidence that mating preferences are altered by changes in the frequency of the various types in the populations. This interesting property has, however, been discovered in a number of instances with several species of *Drosophila*. Usually the change of preference is in the direction of increased mating success for the minority members of the population.

Petit[147,148] has shown that carriers of some of the mutants of *Drosophila melanogaster* behave in this fashion. Assuming that the genotype of the females did not affect the crosses, she determined the proportional insemination of females by males of the wild and mutant type. Under these scoring conditions, males carrying the dominant mutant *Bar* were always at a disadvantage to wild-type males, but the disadvantage was far more pronounced when *Bar* males were present at high frequency. On the other hand, mutant *white* males mated more successfully than wild competitors when *white* males were present in low or in high proportions. At frequencies between 0·4 and 0·8, *white* males were less successful (Fig. 5). It is clear that if all other components of selection were equal, there would be a point of stable equilibrium for this system with *white* at a frequency of 0·4. The other equilibrium point at 0·8 *white* would be unstable, since in

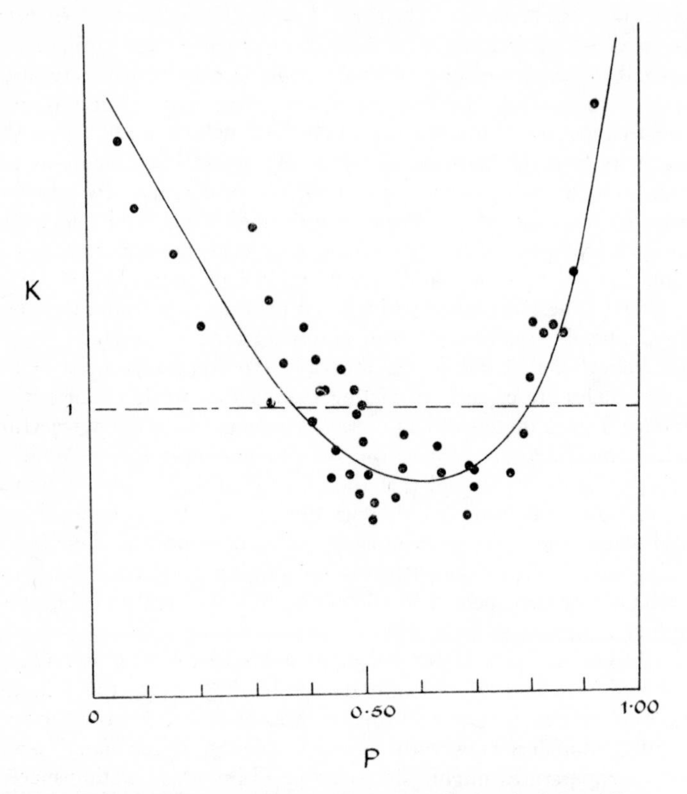

FIG. 5. Variation in the mating success of male *Drosophila melanogaster* carrying the mutant gene *white*. P = gene frequency of *white*;

$$K = \frac{\text{no. of ♀♀ inseminated by } \textit{white} \text{ ♂♂/no. of } \textit{white} \text{ ♂♂}}{\text{no. of ♀♀ inseminated by wild ♂♂/no. of wild ♂♂}}.$$

Values of K greater than 1 indicate an advantage to *white*; values less than 1 indicate an advantage to the wild type. (After Petit[149].)

populations with higher frequencies of *white* the mutant allele would proceed to fixation.

An independent discovery of frequency-dependent mating success in *Drosophila pseudoobscura* was made by Ehrman *et al.*[54] during a behavioural analysis of stocks which had undergone selection for differential geotaxis. The original stocks carried both Arrowhead (AR) and Chiricahua (CH) gene arrangements, and during the course of the selection experiments it became obvious that mating was not random with respect to karyotype. Using the Elens–Wattiaux chamber,[56] which permits direct counts of all classes of matings, Ehrman and her colleagues tested the mating behaviour of AR and CH stocks with or without an initial selection for positive or negative geotaxis.

In both selected and unselected flies there was a slight tendency toward homogametic matings, provided that equal numbers of AR and CH flies were placed in the chamber. On the other hand, if 20 pairs of one type and 5 pairs of the other were introduced, then in many cases the minority group enjoyed an advantage in mating. In the unselected stocks both males and females of each stock of each karyotype showed a very significant effect in every test.

Subsequent work has shown that the mating advantage of minority types in *Drosophila* occurs frequently, though not invariably, with many kinds of differences. The effect has been found in *D. pseudoobscura* from different localities,[54] in *D. persimilis* from different localities or raised at different temperatures,[179] and in three members of the *D. willistoni* group, each species represented by populations from different places.[53]

The basis for the minority advantage remains rather obscure. It is found much more often with males than with females. The effect can be extinguished by piping into the mating chamber olfactory stimuli from flies of the same kind as the minority[52] or by placing males (but not females) of the minority type in an adjacent chamber separated by two layers of gauze with a space between.[55] What the olfactory differences between karyotypes or flies with different developmental temperatures might be is unknown. Whatever the mechanism may be, however, it is clear that the minority advantage represents a powerful frequency-dependent component of selection in *Drosophila*.

Predator-prey Relationships

Another system which is potentially capable of generating frequency-dependent selection is the interaction between a predator and its prey. The effects may take place at the level of interspecific competition between alternate prey species or among the polymorphic forms of a single species.

In what is perhaps the most important paper to appear on the subject, Clarke[18] has examined the available evidence on the frequency-dependent effects of predation giving particular attention to the re-interpretation of neglected data in the literature. He has cited, for example, the work of Reighard[157] on the predatory behaviour of the grey snapper, *Lutianus griseus*, towards the silverside, *Atherina laticeps*.

In an attempt to show that the coloration of coral reef fishes cannot be the product of natural selection, Reighard first wished to establish that the predatory snappers are capable of distinguishing colours and can form and retain associations between colour and palatability. He artificially coloured large numbers of silversides either red or blue and determined that the snappers would readily accept either colour. Then after a period of feeding only blue fish, he offered mixed groups of 5 blue and 5 red fish. Between each successive pair of mixed feedings, blue fish alone were offered. By scoring the order in which fish of the mixed groups were taken, Reighard showed that the snappers are capable of distinguishing between the two colours. The data from one such experiment are shown in table 6. In further tests he demonstrated that the snappers can learn to associate different colours with differences in palatability, retaining the associations for considerable periods of time.

TABLE 6

Predation by the grey snapper (Lutianus griseus) *on artificially coloured silversides* (Atherina laticeps). *Trials were conducted with 5 blue and 5 red fish at a time, except for the last trial which used 4 of each. The colours of the first 5 fish taken in each trial are compared with those of the last 5 taken.* (*After Reighard*[157].)

	Total taken: First five places	Total taken: Last five places	Grand total
Blue	47	7	54
Red	8	46	54
	55	53	108

After showing that the necessary conditions for warning coloration actually occur in predator-prey systems in fishes, Reighard nevertheless argues against the importance of warning coloration. In the specific case of the reef fishes, where many of the brightly-coloured forms are demonstrably palatable, his point is well taken. On the other hand, the general argument against warning coloration seems to have very little to do with the evidence presented.

Clarke[18] has pointed out, however, that the really interesting phenomenon in Reighard's data is the tendency of the predators to take prey of the types to which they are accustomed. The behaviour of the snappers indicates that they will take most rapidly whichever form is commonest in the population. Hence common types will be predated proportionally more heavily than rare ones.

Clarke was able to make a more elegant analysis of a similar phenomenon from Popham's[152,153] data on the predation of a corixid bug, *Sigara distincta*, by the rudd, *Scardinius eryophthalmus*. *Sigara* is capable of altering its colour during development to produce a variety of shades of brown matching its environmental background. Popham showed, as one might expect, that when equal numbers of two colours of prey were offered against a uniform background, the animals which matched the background most closely were least heavily predated. This simple example of protective coloration became more complicated

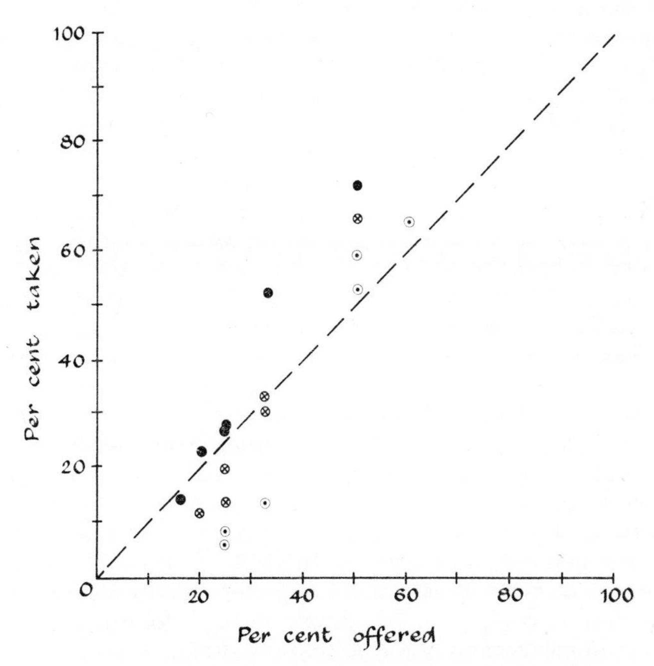

FIG. 6. The proportional predation by the rudd, *Scardinius eryophthalmus*, on three colours of the corixid bug, *Sigara distincta*. Solid circles represent the least cryptic type; circles with dots, the most cryptic. Each type is taken more frequently than expected when common, and less frequently than expected when rare. (After Clarke[18].)

when the proportions of the colours were varied. Although in the two-colour experiments the cryptically coloured insects were always at an advantage, this advantage was most marked when they were at low frequency.

A new characteristic appeared in Popham's experiments when a third colour type was added to the mixture of prey. Even though the third type was least cryptic of all, its effect was to make all three colours disadvantageous when common and advantageous when rare. Thus the most cryptic form was taken more frequently when common and less frequently when rare than its representation in the mixed sample of prey. This relationship is illustrated in Fig. 6.

The principle of frequency-dependent selection by predators may be extended to other vertebrate predators and to inter-specific competition through the work of L. Tinbergen[187] on insect larvae in pine forests. Species present at low densities were found to be taken less often and those at high densities more often than expected from their frequencies in the joint population. Tinbergen suggests that success by a predator in capturing edible prey results in a search for more of the same kind of prey. Further success reinforces the 'searching image' by which the predator recognizes prey. With predators which hunt primarily by sight this image would of course be visual, although olfactory and auditory 'images' would result in similar behaviour.

Clarke[18,209] has raised the question of whether frequency-dependent selection by predators, which he has termed *apostatic selection*, may not play a role in the polymorphism of land snails of the genus *Cepaea*. The two species inhabiting Britain and Western Europe, *Cepaea nemoralis* and *C. hortensis*, display a remarkable range of parallel variations in the colour and pattern of the shell. Individuals may be brown, pink or yellow, with the alleles for colour showing a descending hierarchy of dominance in that order. The shell may bear up to five (or rarely more) dark bands, the presence or absence of bands being controlled by a locus tightly linked to the colour locus. Other loci control variations of the banding pattern, presence or absence of colour in the bands and/or lip of the shell, and a number of other characters. The genetic systems of *C. nemoralis* and *C. hortensis* are largely homologous, although the frequencies of alleles in populations of the two species are often very different. In most localities, for example, the allele for dark lip is fixed in *nemoralis* and that for white lip is fixed in *hortensis*.

In a paper which definitely established that natural selection, operating by means of predators, influences the morph-frequencies in *C. nemoralis*, Cain and Sheppard[14] considered the possibility that frequency-dependent selection might be involved in the maintenance of the polymorphism. They rejected the idea for want of evidence, also citing the negative evidence that in accumulations of shells broken by an important predator, the thrush *Turdus ericetorum*, there did not appear to be the daily fluctuations of frequency of common phenotypes that might be expected if a bird's initial captures affected its subsequent searching behaviour during that day.

Clarke[19] has argued that frequency-dependent selection in the *Cepaea* polymorphism should be detectable in its operation at the inter-specific level. It is unlikely that the selective value of the morphs, apart from the visual properties, will be the same in the two species. It is clear, for example, that in most English colonies, pink and brown shells are much more common in *C. nemoralis* than in *C. hortensis*. If for reasons unrelated to the visual characters one morph is common in one species in a mixed colony, then frequency-dependent predation might be expected to depress the frequency of that morph in the other species. Frequency-dependent selection might therefore be expected to lead to negative correlations of the frequencies of similar morphs in the two species when they occur together in mixed populations.

Lamotte[106] has described some of the most dramatic cases of negative associations of phenotypes in the two species. For example, his sample from Saint Thois consists entirely of pink *nemoralis* (92 individuals) and yellow *hortensis* (23 individuals). Other samples show a preponderance of unbanded and midbanded forms in *nemoralis* and high frequencies of the five-banded form in *hortensis*. Lamotte describes this condition as the 'phenomenon of complementarity', without explanation, offering the data as evidence that selection by predators does not produce the expected parallel variation in the two species.

Clarke[19] has argued that this complementarity is the result of frequency-dependent selection by predators that fail to distinguish between the two species, as indeed did most malacologists until the middle of the nineteenth century. Fig. 7 summarizes his data from mixed colonies in an area within sixty miles of Oxford. The scatter-diagram shows the relationship between the proportion of yellow shells in *nemoralis* and the proportion of yellow effectively unbanded shells in *hortensis*. The latter class includes all yellow shells either with the upper bands absent or with pigmentless bands. From the previous work of Cain and Sheppard[14] and Clarke[17] on the different responses of the two species to visual predation, the diagram might be expected to show a positive correlation between these two classes. On the contrary, there is not only no overall correlation of any sort, but there is also a *negative* correlation within each of the two general classes of habitats (open habitats: $r = -0{\cdot}66$, $P < 0{\cdot}001$; woodland habitats: $r = -0{\cdot}474$, $P < 0{\cdot}1$, n.s.).

In order to investigate the behaviour of wild birds under more

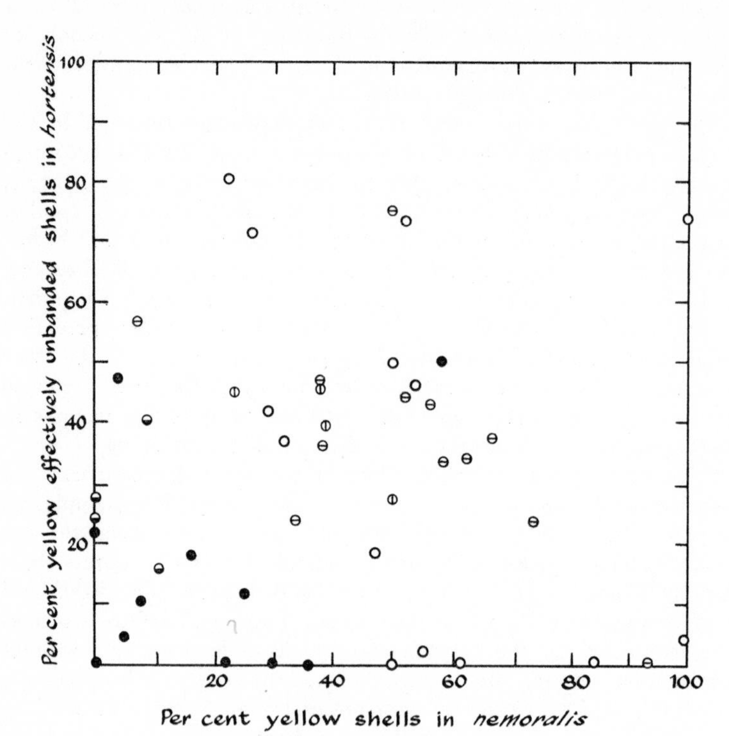

KEY

- ● BEECHWOODS
- ◒ OTHER DECIDUOUS WOODS
- ⊖ ROUGH HERBAGE
- ⦶ FENS
- ○ GRASSLANDS

FIG. 7. The relationship between the proportion of effectively unbanded shells (*i.e.* with bands 1 and 2 absent) in *Cepaea hortensis* and the proportion of yellow shells in *C. nemoralis* from samples in which both species are represented. (After Clarke[19].)

controlled conditions, Allen and Clarke[1] have observed the 'predation' of artificial prey in experiments where the proportion of two types of prey could be varied. The baits, made of flour and lard dyed green or brown with food colouring, were displayed on a grid laid out on a green lawn. For the first seven days a random array of 90% green and 10% brown was offered. After a two-day interval, the proportions were reversed for another seven-day period. Blackbirds (*Turdus merula*), House Sparrows (*Passer domesticus*), Dunnocks (*Prunella modularis*), and Starlings (*Sturnus vulgaris*) took sufficient numbers of the baits for analysis. Comparing the proportions of colours taken, with the proportions expected on the basis of constant selective values over the entire experiment, Allen and Clarke were able to show that all four species in both parts of the experiment took fewer of the rare type and more of the common type than expected. It appears, therefore, that these birds are predating in a frequency-dependent manner and that they are capable of reversing the direction of their selection in a very short time.

TABLE 7

Numbers of baits of different colours taken during individual visits by Blackbirds (Turdus merula) *on a single day. (After Allen and Clarke*[1]*.)*

Visit	Number of Baits Taken Green	Brown
1	4	0
2	7	0
3	3	0
4	3	0
5	9	0
6	5	0
7	5	0
8	2	0
9	0	10
10	24	0
11	3	0
12	5	0
13	1	0
14	0	12
Total	71	22

Of even more interest is the indication that individual birds do continue to look for food of a kind recently taken, as the concept of a 'searching image' predicts. Table 7 shows the records of individual visits by Blackbirds on a single day during the display of 90% green. The majority of the birds took only greens; but it is obvious that on the two occasions when browns were taken, the bird continued to search for brown baits even though these were scarce. In each of these two visits, the birds took at least half of all the brown baits on the whole grid without picking up a single green.

A final example of the operation of frequency-dependent selection in predator-prey relationships may be drawn from one of the classical examples of natural selection, Batesian mimicry. Fisher[62] clearly pointed out the basic distinction between the modes of selection in Mullerian mimicry and in Batesian mimicry. In Mullerian mimicry, since there is selection for uniformity and no theoretical ceiling to population numbers, there will be directional selection which is largely frequency-independent in its operation. In Batesian mimicry, on the other hand, there is necessarily an upper limit to the number of mimics which a given model will support. Although Brower[9] has shown that this limit is not so restrictive as was once supposed and that it depends also on the degree of unpleasantness of the model, nevertheless a ceiling will inevitably be reached when mimics are so common that predators do not gain experience with the model. Thus the selective advantage of the mimic is frequency-dependent, declining with increasing abundance; and the result will often be a stable polymorphism. The most extravagant example of such a development is the remarkable variety of forms of the African swallowtail butterfly, *Papilio dardanus*.

Although the theoretical case for the importance of frequency-dependent selection in the maintenance of mimetic polymorphism is quite convincing, the data bearing directly on this point are rather meagre. Using some of Carpenter's collections, Sheppard[169] has demonstrated a relationship between the frequency of five mimetic forms of *Pseudacrea eurytus* and that of the five corresponding species of *Bematistes* which serve as models. The correlation between them is very high ($r = 0{\cdot}92$). Such a correspondence will not always be so dramatic, as Sheppard has pointed out in *Papilio dardanus*. The important numerical relationship is the ratio of the total number of mimics to models. If more than one species is mimicking the same model then the system is one of balanced competition as Clarke[18] has termed it. In such a system the influence of the model on the frequency of a relatively rare mimic, such as *P. dardanus*, will be mediated by the joint frequency of more common mimics of the same model. In this case *Pseudacrea eurytus*, being always more common than all other mimics, exerts such a control.

Among the butterflies of the eastern United States, there is evidence for frequency-dependent selection in *Papilio glaucus*, the tiger swallowtail. The females of this species are of two major types, one resembling the monomorphic yellow and black males and the other with the yellow replaced by a dusky brown or black. The black females belong to a mimetic complex dependent on the distasteful model *Battus philenor*.[8] A number of other species also share the model: *Papilio troilus*, *P. polyxenes*, *Limenitis arthemis astyanax* and probably also the female of *Speyeria diana*.

The polymorphism of *P. glaucus* shows a rather unusual type of genetic control which has a direct bearing on its population dynamics. Clarke and Sheppard[26] have shown that females, with very rare exceptions, produce female offspring which are like themselves. This method of inheritance probably implies Y linkage of the locus, the female in butterflies being the heterogametic sex. Cytoplasmic inheritance is unlikely since occasional females show the yellow coloration on one side and the black on the other. If the linkage in Y is the correct explanation, then it follows that the polymorphism cannot be maintained by heterozygote advantage, and the two Y-determined forms will be in direct competition with each other. While this in itself might be considered as indirect evidence for a frequency-dependent component to balance the system, there is additional evidence that the frequency of the black form is determined by the abundance of the model *Battus philenor*. The range of *P. glaucus* exceeds that of *B. philenor* both to the north and to the south. In those areas where *philenor* is common, the black form of *glaucus* predominates; while outside the range of *philenor*, the frequency of black *glaucus* falls off sharply. The fact that black *glaucus* females are rare both north and south of the range of *philenor* argues that the polymorphism is not the result of climatic factors. The available data on the joint distribution of *philenor* and *glaucus* in local populations[10] are as yet insufficient to allow more than a general statement on distribution. Here is a system in which more random samples from local populations are badly needed.

An experimental approach to the study of frequency-dependent selection in mimicry has recently been initiated by O'Donald and Pilecki.[143] Using coloured baits of flour and lard, they displayed randomly placed assortments consisting of 20% edible yellows, 25% green models, 25% blue models, and 30% edible mimics (either 25% green and 5% blue or *vice versa*). In two experiments the degree of distastefulness of the models was varied by treating them with solutions of quinine hydrochloride of two different strengths. The predators in all cases were a flock of House Sparrows (*Passer domesticus*).

As might be expected, the edible non-mimetic baits were taken much more readily than the models and mimics, a clear demonstration of the selective value of mimicry. There was no difference in the predation on models and mimics of any one colour. The significant point, however, is that in the experiment with moderately distasteful models, the colour with the smaller proportion of mimics was taken significantly less often than expected on the basis of its frequency in the total population of models and mimics. Thus the rare mimic enjoyed a selective advantage. Curiously enough, the effect disappeared when models were made more disagreeable. It seems that in this system a sufficiently noxious model will protect a fairly common mimic to the same degree as a rare one. O'Donald and Pilecki suggest, therefore, that with extremely distasteful models, mimetic polymorphism is unlikely to evolve. Further work on the relationships of the degree of distastefulness of models, the proportions of models and mimics, and the perfection of the mimetic patterns should provide more understanding of the dynamics of frequency-dependent selection in mimicry.

Parasitism and Disease

A system which is analogous to the predator-prey relationship is found in the reaction of organisms to the pressures of parasitism and disease. In a prescient and thought-provoking paper, Haldane[78] has pointed out the manner in which disease may influence the course of evolution. One of a number of novel ideas which he presents is the possibility that pathogens may produce frequency-dependent selection favouring rare biochemical variants. He postulates that the great biochemical diversity of populations comes about as a result of differing resistance to strains of disease organisms. Haldane's thesis even provides a plausible explanation for the partial failure to detect a relationship between blood groups and disease. The proper approach is, as Haldane points out, to look for such associations within epidemics caused by a single strain of the disease organism, not over the whole range of hosts and pathogens.

Perhaps the strongest summary of evidence in favour of this diversifying effect of parasites and disease organisms has been put together by Damian[36] in his discussion of the phenomenon of *molecular mimicry*. It has been known for many years that human pathogens and parasites frequently contain antigens similar or identical to those found in the host. Although the source of these antigens has been hotly debated, it now seems to be well established that in many cases they are produced by the parasite itself. Thus antigens related to those of the ABO blood groups are produced by a wide variety of such organisms as *Ascaris lumbricoides*, *Necator americanus*, *Taenia solium*, *Fasciola hepatica*, *Schistosoma mansoni*, and many of the gram-negative bacteria.

The most reasonable explanation for the occurrence of these shared antigens lies in the opportunity they afford for the circumvention of the hosts' immunological defences. The host will be unable to recognize the parasite as non-self, and hence unable to manufacture antibodies to the antigen. It follows therefore that hosts with rare antigenic phenotypes will be less susceptible to infection by parasites and pathogens. The advantage will be frequency-dependent, however, since new infectious strains may be expected to develop in response to new challenges. Haldane has cited the case of wheat rust as a model for this type of system. The development of new varieties of wheat resistant to the current strains of *Puccinia graminis* is inevitably followed by the appearance of new strains of rust. The unnatural density of populations in agricultural practice speeds up the time-sequence of the system.

Damian has made the point that the adaptation of parasites by way of molecular mimicry makes it likely that heterozygotes will be selectively inferior to homozygotes. The conditions which allow persons of AB blood type to be 'universal recipients' also prevent them from reacting to the presence of two mimetic types, a double disadvantage. This inferiority of the heterozygote raises a serious and unresolved problem for the hypothesis. It is not at all clear how new forms, appearing first in heterozygous conditions, are able to become established in the populations.

A dramatic demonstration of the role of molecular mimicry in the adaptation of pathogens to their hosts has been made by Simonsen and Harris[174] using Rous-1 virus. Normally the host of this virus is the chicken; turkeys are resistant. In these experiments turkeys were made immunologically tolerant to chicken blood, and in the course of the conversion they also became sensitive to the Rous-1 virus. The most probable interpretation of these results is that the virus mimics an antigen of its normal host. When the turkeys acquire tolerance to the chicken antigens they therefore also lose the ability to react to virus antigen. Here again is a model for frequency-dependent selection for biochemical diversity, since by reversing the argument, chickens with antigenic variants would acquire immunity to the virus.

A possible complication of this straightforward interpretation has been introduced by Schad.[163] He has suggested additional reasons for antigenic resemblances of two parasites of the same host. That the similarities may be significant in interspecific competition within the host is suggested by the occurrence of non-reciprocal cross reactions. Schad cites examples in roundworms, tapeworms and flukes in which infection by a parasite results in an immunization of the host against another species of parasite, but not against the original parasite itself. It is not clear how such cross reactions can be effected, but in any case the interspecific competition does not explain the initial similarity of one or the other of the parasites to the host.

Competition

There is a large and growing number of examples in the literature in which frequency-dependent selection has been demonstrated without the identification of any specific agent. It is convenient to group such cases together operationally, treating them as examples of competition. This, of course, does not set them apart, since selection inevitably implies intraspecific competition; but the category is useful in that it is through the observation of competition that we are able to detect the action of frequency-dependent selection.

The early work of Dobzhansky and his collaborators[42,108] with population cages of *Drosophila* has made it quite clear that the selective value of a genotype depends on the other genotypes in the population and that the interactions are not at all straightforward. From observations of mixtures of the inversion types Standard (ST), Chiricahua (CH), and Arrowhead (AR) they have shown that one cannot predict the selective values of untested genotypes in combination. For example, one cannot predict the values of CH/CH, CH/AR, and AR/AR in combination, from the behaviour of CH and AR when tested singly with ST.

Lewontin[111] and Spiess[178] have extended these observations to show that the selective values of genotypes depend not only on the identity of other genotypes in the population but also on their frequency. Lewontin chose 22 strains of *Drosophila melanogaster* and compared their larval viability both in pure and in mixed cultures. The mixed culture consisted of equal numbers of larvae of the test strain and a strain carrying the mutant white. In some of the strains examined the viability was higher in mixed than in pure culture. Thus fitness increased as frequency decreased. Lewontin has termed this relationship 'genetic facilitation'.

In Spiess'[178] experiments, various combinations of chromosome arrangements in *Drosophila persimilis* were tested in population cages. Combined with the Whitney (WH) arrangement of the third chromosome, both Klamath (KL) and Mendocino (MD) reached stable polymorphisms with a frequency for WH of 65–70%. Spiess obtained evidence that during the approach to equilibrium the selective values of the genotypes changed, since no single set of selective values sufficed to describe the changes of chromosome frequency in either set of populations.

Spiess suggested three possible explanations of the altered selective values. Either there had been a readjustment of the residual genome to produce new interactions, or there had been a change in the population-cage environment, or else the selective values were frequency-dependent. The first hypothesis seems unlikely as the behaviour of these systems was determinate, and it is inherently improbable that the same event would occur each time. Also since runs were made at different times, it is unlikely that environmental variables in the population cages would account for the change. Hence Spiess argues that the outcome of the competition is determined in these cases by the frequency-dependence of the selective values. Interestingly enough, when KL and MD, both rare in nature, were tested together the results were indeterminate and unpredictable from their behaviour with WH.

Sokal and his colleagues[176,177] have shown, in more detail, the relationship of selective value and frequency in laboratory cultures of *Tribolium castaneum*. Using different strains of *Tribolium* marked with the mutants *sooty* or *black*, they have reared mixtures of mutant and wild-type beetles from egg to adult, with different initial densities and varying frequencies of the types. The genotypes were mixed in Hardy–Weinberg proportions. As one might expect, they found that the performance of a type in mixed culture was not predictable from the pure cultures. Usually there was an enhancement of survival of all three genotypes, with the values varying with both frequency and density. The relationships cannot be simply described, but the most noticeable characteristic is a marked enhancement of the survivorship of the $+/+$ homozygote when the *black* gene is at high frequency. The reverse relationship holds at low density, but the b/b homozygote is especially unsuccessful at high density when $+/+$ is at high frequency. That these frequency-dependent relationships may lead to polymorphism is shown by selection studies in laboratory populations; on the other hand polymorphism in wild populations is unknown.

A similar study in *Drosophila melanogaster* has been carried out by Kojima and Yarbrough on the Esterase 6 locus.[103,207] Using a population which was polymorphic for the fast and slow alleles of Esterase 6, they studied both larval viability and the performance of the alleles in population cages. They found that when groups of fertilized eggs representing different ratios of the fast and slow alleles in Hardy–Weinberg proportions were allowed to develop, the adult ratios indicated selection in favour of whichever allele was deficient compared with the original equilibrium value of 0·3 for Esterase 6 F. In addition they attempted to describe the observed changes in population cages by fitting selective values iteratively on a computer. The results seemed to indicate that a better fit was possible if selective values were allowed to vary with frequency. Kojima's group has also noted similar effects with polymorphisms for chromosome inversions in *Drosophila ananassae*[188] and for an alcohol dehydrogenase in *D. melanogaster*.[102]

One of the chief difficulties in the interpretation of the results of population-cage experiments is the problem of distinguishing between the change of selective values with time and the change of selective value with change of gene frequency. Harding, Allard, and Smeltzer[81] have devised a technique for making the distinction in a population of lima beans, *Phaseolus lunatus*. Lima beans are characterized by a high frequency of self-fertilization; but if a population is continued, the proportion of heterozygotes approaches an equilibrium value of about 10% after five generations. Using the S locus for seed-coat pattern, Harding *et al.* have tested the alternate hypotheses of (1) the evolution of heterozygote advantage over time or (2) the dependence of fitness on frequency. To do this they have used populations synthesized at two different times from a pair of highly inbred lines. These provided contemporaneous stocks from the early generations (F_4 or F_5) and the late generations (F_{13} or F_{14}) of the same kind of cross. One set of experimental populations was set up entirely from the late stocks by varying the proportions of homozygotes and heterozygotes. A second set of populations was constructed also with varying proportions of genotypes but with the homozygotes drawn from early and the heterozygotes from late stocks and *vice versa*. Since the results from all populations were essentially the same, the possibility of a change in fitness with time can be excluded. All the populations showed a negative correlation between fitness and frequency, the heterozygotes being at a great advantage when very rare. The effect appears to be the result of competitive interactions between neighbouring plants.

Disruptive Selection

Selection in natural populations may conform to any of three distinctly different modes of operation. Historically, two of these have commanded the principal attention of population geneticists. On the one hand, the concept of stabilizing selection embodies the response of a population to forces producing detrimental deviations from the optimal phenotype. On the other, directional selection describes the process whereby the population changes with time to incorporate new, selectively favoured alleles into the genetic system.

The third mode, disruptive selection, attracted relatively little attention until it was discussed by Mather[120] in a paper which has become a classic. If several optimal phenotypes are favoured by selection, with intermediates between them having a lower fitness, a number of interesting possibilities arise. These have been explored in some detail by Mather,[120] Thoday and his colleagues (see [71] for references), and Maynard Smith.[121,122] All of these authors have reached the conclusion that disruptive selection may in certain circumstances lead to the development of either stable polymorphism or reproductive isolation.

The case of simple disruptive selection with constant selective values in a uniform environment can be dealt with very briefly. Li[115] has shown that selection against the heterozygote in such a system yields a single point of unstable equilibrium with any displacement from that point leading to the fixation of one of the alternative alleles. Dominance simply removes the equilibrium point and makes the mode of selection identical to the case of directional selection. My conclusion is therefore that disruptive selection *per se* is neither a necessary nor a sufficient condition for the production or maintenance of stable polymorphism or for the development of reproductive isolation

This is not to say that disruptive selection does not play an important role in both these processes, and in others as well. Its action in polymorphism may be inferred from a substantial body of data from the field, although its effect on the origin of isolation remains only a possibility. It is convenient to make a distinction here between the maintenance of a polymorphism or the way in which the forms are prevented from being lost from the population, and the *dissociation* of the forms or the way in which their distinctiveness is developed and preserved. Dissociation would not be expected to operate in the case of elementary biochemical polymorphism, but it should be a feature of many complex polymorphisms, in which multiple elements are controlled by a single segregating locus.

Mather[120] has provided the clue to the role of disruptive selection in polymorphism. He speaks of Batesian mimicry as a 'situation where a number of different optimal phenotypes would be advantageous, the phenotypes being tied together in that the success of each would depend on the frequency of the others, including non-mimetic phenotypes.' Disruptive selection is therefore responsible for the dissociation of forms while the maintenance of the polymorphism (the 'tie') is the result of frequency-dependent selection. It is a unique property of frequency-dependent selection, the ability to maintain a polymorphism even in the extreme case of heterozygote disadvantage, which enables it to stabilize a system involving disruptive selection.

An examination of the experimental work by Thoday and his colleagues reveals that in every case frequency-dependent selection is built into the design of the experiment. Although the details of the mating system vary from one experiment to the next, the essential points are the same in each. Stocks of *Drosophila melanogaster* were subjected to disruptive selection for the number of sternopleural chaetae. From each generation, flies with high and low numbers of chaetae were chosen as parents, excluding those with intermediate numbers. Divergence of the high and low lines occurred with surprising rapidity, leading in some cases to polymorphism and in others to incipient reproductive isolation. In each generation, however, equal numbers of flies with high and low values were chosen. Thus if flies with high numbers of chaetae are rare, a greater proportion of them will be chosen as parents and *vice versa*. The phenomenon which is under test here is not the maintenance of a polymorphism but the dissociation of morphs. Since the dissociation is successful, frequency-dependent selection maintains the polymorphism, or the balanced competition in the case of incipient isolation.

Thoday has pointed out (personal communication) that equality of numbers is not essential to the progress of the dissociation. It is sufficient for frequency-dependent selection that enough of each are selected to prevent swamping of the incipient differences by the more numerous class.

Maynard Smith[122] has investigated mathematically the possibility that disruptive selection with constant selective values in a heterogeneous environment may lead to a polymorphism. He has shown that there are conditions in which stable equilibria may be attained in such a system, but that they are rather restrictive. Specifically the population sizes in the different portions of the environment must be subject to separate density-dependent regulation, and the selective coefficients must be of the order of 30%. While neither condition is particularly unlikely, it would probably be more realistic to postulate differential density-dependent effects on the different phenotypes, *i.e.* frequency-dependent selection. In any case this model does not describe the relationships of Thoday's experiments. The extra component of frequency-dependent selection no doubt explains the ease with which the dissociation was achieved and maintained.

Another model employing fixed selective values, proposed originally by Levene,[107] and recently discussed by Levins and MacArthur,[110] also leads to polymorphism in a heterogeneous environment. As Maynard Smith[121] has shown, however, this regime depends on average heterosis over all the environmental patches.

The final problem of some importance in cases of disruptive selection is that of the outcome of the process. In Mather's[120] original discussion he foresaw the possibility of the development of reproductive isolation, provided that the optimal phenotypes are distinct and independent and that the environmental conditions are persistent. This is the model of sympatric speciation which Maynard Smith has made explicit. On the other hand, Mather implied that if the forms are mutually interdependent, *i.e.* if selection is frequency-dependent, then the outcome will depend on the relative magnitude of this interdependence and of competition. The essence of the problem lies in the manner in which the selection against heterozygotes is resolved. There are two possible ways in which their recurrent selective elimination may be minimized: either the production of heterozygotes may be reduced through selection for homogamy,[18,122] or the expression of heterozygotes may be modified through the evolution of dominance.[20,25] Which of these results finally obtains will depend both on the genetic make-up of the species involved and on the nature of the environment. These processes will be discussed in more detail in the ensuing chapters.

Conclusions

Frequency-dependent selection as a mechanism promoting genetic diversity has received relatively little attention in classical population genetics. As a system which permits the maintenance of polymorphism with selective neutrality at the point of equilibrium, frequency-dependent selection may avoid the problem of excessive genetic load in such cases.

Many familiar biological systems involve frequency-dependent selection. Common examples are non-random mating especially with respect to outbreeding mechanisms, predator-prey relationships, parasitism, and disease. Frequency-dependent selection may be detected in competitive situations where the agent of selection cannot be identified. It also may be expected when disruptive selection is operating, and may lead to the development of either dominance or reproductive isolation.

4 : The Evolution of Dominance

'If the substitution of mutant for primitive genes has played any part in evolution these observations require that the wild allelomorphs must *become* dominant to their unsuccessful competitors.' R. A. Fisher: *The Genetical Theory of Natural Selection.*[62]

In the preceding chapter we have seen that frequency-dependent selection has a number of intriguing properties. Not the least of these is the ability to maintain a polymorphism in which the heterozygote is selectively inferior to both homozygotes. Although such a polymorphism may be stable, it nevertheless entails a continuing loss of individuals in each generation, *i.e.* a genetic load. An escape, however, from this consequence is possible provided that the dominance relationships of the alleles may be modified.

The problem of the evolution of dominance under frequency-dependent selection is part of the larger question of the modification of dominance in any genetic system. This question has been the topic of a recurring debate for the past forty years, the most recent round being initiated by Crosby in 1963.[32]

Fisher's Theory

Following the excesses of the period of Bateson's presence-and-absence theory, the first serious attempt to account for the phenomenon of dominance was made by R. A. Fisher[60] who proposed his theory of dominance modification primarily to account for the initially surprising observation that alleles of the 'wild type' almost invariably are dominant to mutants derived from them. On the other hand, mutant alleles, such as the white-eye series in *Drosophila melanogaster* and the albino series in rats,[201] usually show no dominance among themselves, but in the heterozygous condition produce phenotypes intermediate between the two homozygotes.

Fisher assumed that mutants, when they first appeared, would generally show intermediacy in the heterozygous condition. If the effect of a mutant were to lower the fitness of the organism, then selection would favour those individuals in which the heterozygous expression of the mutant was reduced, and dominance of the wild type would evolve. Recurrent mutation would permit the slow evolution of dominance over periods of time which could even be longer than the lifetime of species, since homologous mutations are characteristic of related species. Initially only the heterozygotes would be subject to selective modification since mutant homozygotes for deleterious genes would be exceedingly rare. After the achievement of complete dominance, however, the frequency of the mutant might rise to the point at which modification of the mutant homozygote would become possible. Not all mutants could undergo the evolutionary modification of dominance. Mutants producing very severe defects or lethality in the heterozygous condition would be subject to such intense counter-selection that they would be eliminated before exerting an appreciable effect on the gene pool. Even with a mutant that only mildly handicapped the heterozygote, initial progress would be slow. The process would, however, accelerate as some measure of dominance was attained.

In his initial paper Fisher[60] cited other indirect evidence favouring his hypothesis. He pointed out that when some mutant genes in *Drosophila* (*e.g. eyeless*[133]) are maintained in homozygous condition in laboratory stocks, the expression of the mutant is gradually reduced. Since the full expression is regained upon outcrossing and re-extraction of the mutant, the change is obviously not in the allele itself but in the genetic background. This result implies that the type of variation required for dominance modification actually exists in populations.

Later in the same year, Fisher[61] introduced two more examples which have been widely discussed since that time. He cited the behaviour of *crinkled dwarf*, a recessive mutation of cotton. In crosses between species of cotton the dominance relationships break down, although this example now seems to be much more complex than it appeared at the time.[66,88] The second case is of particular interest because it involves apparent exceptions to the rule that mutants are usually recessive to the wild type. In the domestic chicken there are a number of traits such as *Crest*, *Rose Comb*, and *Barred* which are dominant to their wild type alleles. For an explanation Fisher invoked the circumstances which may have attended the domestication of the jungle fowl in eastern Asia. He suggested that conditions were probably much the same as they are today, with considerable outbreeding of domestic flocks to wild cocks. Any attempts to retain unusual variations within the flocks, a characteristically human endeavour, would produce selection for dominance, since heterozygotes in which the expression was enhanced would be retained by the owner.

Experimental Modification of Dominance

Soon after the publication of Fisher's theory a number of experiments were designed to investigate the modification of dominance. Fisher himself[64,65] took seven dominant mutations of domestic chickens and bred each one back into wild jungle fowl stocks. After five generations of backcrossing, the heterozygotes were crossed to obtain homozygotes for comparison. The mutants responded in different degrees to the treatment. *Dominant White* (*Pile*) retained a large measure of dominance. *Feathered Feet* became nearly recessive, while most of the rest gave intermediate heterozygotes, easily distinguished from homozygous mutants. In the case of *Crest*, Fisher found that the homozygote in the genetic environment of the wild stock produced a cerebral hernia. This condition is therefore a recessive trait in the wild stock, but it has been entirely suppressed in breeds characterized by the homozygous crested condition. The expression of the character of the crest has therefore been enhanced in domestic breeds while at the same time the accompanying hernia has been reduced.

An experiment by Ford,[69] using the Currant Moth, *Abraxas grossulariata*, has shown the ease with which dominance modification may be accomplished by artificial selection. In the variety *lutea*, the usual pale cream ground colour of the wings is altered to a deep yellow. The heterozygote is intermediate, although there is considerable individual variation in the depth of pigmentation. Beginning with a cross of a heterozygote and a normal white moth, Ford selected the lightest and darkest heterozygotes as the parents for successive generations. As early as the third generation of selection in the light line, a cross between a pale heterozygote and a pure white produced some offspring which were undoubtedly homozygous *lutea*, indicating that *lutea* had become completely recessive in the white individual. In the dark line, the next generation was shown to contain heterozygous *lutea* as dark as the modal class for homozygous *lutea*. The experiment therefore demonstrates quite clearly the availability of genetic variation capable of responding to the pressure of selection for dominance modification.

Fisher and Holt[67] carried out a similar experiment using Danforth's short tail (S^d) in mice. In heterozygous condition this mutant produces a shortened and often deformed tail. The homozygote is completely tailless and dies shortly after birth, with severe defects of the kidneys and an imperforate anus. Fisher and Holt outcrossed the mutant to a variety of laboratory lines and selected for increased or decreased expression in the heterozygote. After two years of selection, the mutant behaved as an almost complete recessive in the positive line, with some of the homozygotes capable of surviving beyond weaning. On the other hand, although progress was achieved in the negative line, natural selection successfully opposed the attainment of complete dominance by S^d.

Traditional Criticisms of Fisher's Theory

The principal questions inherent in Fisher's theory of dominance modification were raised very soon after its initial publication. Wright,[202] in fact, introduced a more explicit model which has served as a basis for most of the subsequent discussion. He proposed the case of a major locus having a pair of alleles (A and a) with an intermediate heterozygote modified by one specific locus. A dominant allele (M) at the modifier locus established complete dominance of one allele at the major locus, while the recessive allele (m) was without effect. The argument was couched in terms of change of gene frequency at the modifier locus.

Wright's chief criticism[202,203] is that the initial stages of the increase in the frequency of the modifier would take place with extreme slowness, the change being of the same order of magnitude as the mutation rate. He agreed with Fisher that once heterozygotes at the major locus had achieved an appreciable frequency, then the system would work, but he could not accept its efficacy during the initial stages. Unless there exists a very large array of genes whose only function is the modification of dominance, a hypothesis that is inherently unlikely because it suggests an infinite regression of modifiers, then the modifier genes must have other primary functions. These effects would be expected to take precedence over the small second-order effects on dominance. Even if a special class of modifier genes should exist, random fluctuations of frequency would impede or halt the first slow stages of selection. Haldane[76] emphasized the very narrow range of fitness values over which selection of modifiers would have to take place. The combination AAMm must have a fitness equal to or less than one, or M would simply be selected on its own merits. On the other hand, the disadvantage of AAMm must be less than the advantage which Mm confers on Aa or else M will be lost by selection in the homozygote. Thus Mm must be very nearly neutral in the combination AAMm.

Both Haldane and Wright suggested that dominance and recessiveness are properties of the allele itself, the more active allele determining the phenotype of the heterozygote. They considered that a kind of 'evolution of dominance' occurred as more active alleles were selected over less active ones. Selection would favour any allele capable of producing sufficient product to yield the normal phenotype even when paired with a defective allele. Haldane[76] also suggested that this increased activity might be achieved as the result of gene duplication.

The original discussion between Fisher and Wright eventually led to an uneasy truce. It was generally accepted that expression of alleles in the heterozygote could be either enhanced or suppressed by selection, but there was no agreement on the likelihood of dominance modification in natural populations or on its evolutionary importance.

Recent Debate

Interest in this traditional problem was reawakened by Crosby's[32] renewed criticism of the Fisher theory, although the points which he raises are by no means original. He discusses the problem again in the context of Wright's[202] model, considering the fate of a single modifier locus. He invokes the traditional dilemma that if AA is sufficiently superior to Aa to bring about selection for the dominance of A, then selection will reduce the frequency of a to such a level that favourable dominance modifiers will be lost by drift. On the other hand, increasing the frequency of heterozygotes, in which the modifiers would be at an advantage, can only occur if selection against the heterozygote is relaxed. Hence the selective advantage of the modifier is also reduced at the same time.

Crosby rejects the experimental results in *Abraxas* and the chicken as irrelevant. In each case selection for dominance takes place in populations containing heterozygotes at frequencies far beyond those expected in the classical model. It is also possible that the selection for suppression of the hernia associated with *Crest* took place in homozygotes rather than heterozygotes.

Crosby's second criticism concerns the nature of dominance. He challenges Fisher's assumption that heterozygotes will initially display an intermediate phenotype. In other words, there would seem to be no reason why activity should be directly related to dosage.

In order to demonstrate the ineffectiveness of selection in modifying dominance Crosby has developed a computer simulation of the Wright model. Unfortunately the relevance of the results is vitiated by limitations on the population size (512) and on the duration of the runs.

Mayo[123] has pointed out that Crosby's work simply emphasizes the fact that the initial progress of modification is very slow. Crosby does not discuss the point, agreed on by Fisher and Wright, that selection becomes increasingly effective once the initial phases of modification have been accomplished. The question remains, therefore, whether with this model sampling effects will cancel out the effects of selection.

A rather different attack on Fisher's conclusions has been made by Ewens.[59] He maintains that the apparently increasing effectiveness of selection as modification proceeds is due to a mathematical error in Fisher's argument. Although Mayo[123] disputes in detail Ewens' alternative mathematical derivation, the essential point is simple. Ewens uses the model, which is basically Wright's and not Fisher's, in which the single dominance modifier is itself dominant. Hence when this modifier approaches fixation, and dominance is essentially established, its recessive alternate allele is at such a low frequency that further selection is ineffective. This is not at all the point at issue. It is in the intermediate range of frequencies that the increase in the efficacy of selection will take place. Furthermore the final slowing down of selection for the modifier only occurs with the specific model in which the modifier is dominant.

The present status of the case for the classical model of dominance modification has been summarized by Sheppard and Ford.[171] They have emphasized once again that dominance is the property of the character in question and not of the allele, citing the example of the sickle-cell polymorphism in which haemoglobin type is intermediate while the associated anaemia is recessive. They have clarified the meaning of experiments on artificial selection for dominance, pointing out that the principal function of studies of this sort is to demonstrate the existence of the necessary genetic variability for selection to act upon. And finally, they have identified the real issue: whether a selective pressure of the order of 10^{-6} per generation is sufficient to bring about an increase in the frequency of modifiers over long periods of time in spite of genetic drift.

O'Donald's Theory

A break with the traditional approach to dominance modification has been proposed by O'Donald.[141] He has abandoned the one-locus model, introduced by Wright in 1929, which has been the centre of so much controversy. Instead he has brought the methods of quantitative genetics to bear on the problem. It is a truism that the expression, and hence the selective value, of any particular allele will depend to some extent on the rest of the accompanying genotype, or the genetic background. The selective coefficient of the allele can therefore be treated as a variable with a mean and variance. Selection for the modification of dominance thus involves the alteration of the mean selective coefficient ($\bar{S}$) of the heterozygote, rather than the change in frequency of a particular allele.

The question of dominance modification thus hinges on the amount of additive genetic variance of the selective coefficient of the allele and on the amount of time available for the operation of selection. O'Donald[141,142] has considered the classical problem of recurrent deleterious mutation, as well as that of the spread of an advantageous gene, which will be discussed later. The number of generations involved in the modification of the expression of a deleterious mutant is so large that a direct computer model exceeds the practical limits of even a very fast computer, but O'Donald[141] has used a mathematical model to simulate thousands of generations in the machine computations. Beginning with the genotypes AA, Aa, and aa with selective coefficients of 0·2, 0·3, and 0·4 respectively and with an additive genetic variance in the selective coefficient of Aa of 0·005, O'Donald has followed the progress of dominance modification over a period of 52,000 generations. Fig. 8 shows the results. The selective coefficient of Aa is slowly reduced over a period of about 48,000 generations until it reaches the value of the AA homozygote. Then, in this model, the selective coefficient decreases dramatically over the next 500 generations to produce over-dominance, while the gene frequency of a increases to establish a balanced polymorphism.

Whether or not over-dominance develops will depend on the nature of the change brought about by the mutant allele a. If it is simply a quantitative deviation from the optimal phenotype AA, then the restoration of the optimum is all that can be achieved. However if the change is a qualitative one, the possibility of a new optimum for the Aa phenotype opens the way for over-dominance to develop.

The real question raised by O'Donald's model is whether these conditions represent a reasonable approximation to those existing in natural populations. In particular, are variances of this magnitude likely to be found associated with the fitness of heterozygotes for newly occurring mutations? Although the data bearing on this point are meagre, O'Donald[142] has attempted to evaluate the case of *Abraxas grossulariata* analyzed by Ford.[69] From Ford's colour scores of the heterozygotes, O'Donald makes a rough estimate that the additive genetic standard deviation is 0·3 times the difference between the heterozygote and the homozygote. If the heterozygote suffered from a loss of fitness of 10% (S = 0·1) and the mutation rate of the *lutea* gene was 10^{-5}, then O'Donald's approximate algebraic solution leads to an estimate of 259,000 generations for the attainment of dominance. The estimated additive genetic variance in fitness in this model would be 0·0009, only about one-fifth that of the example

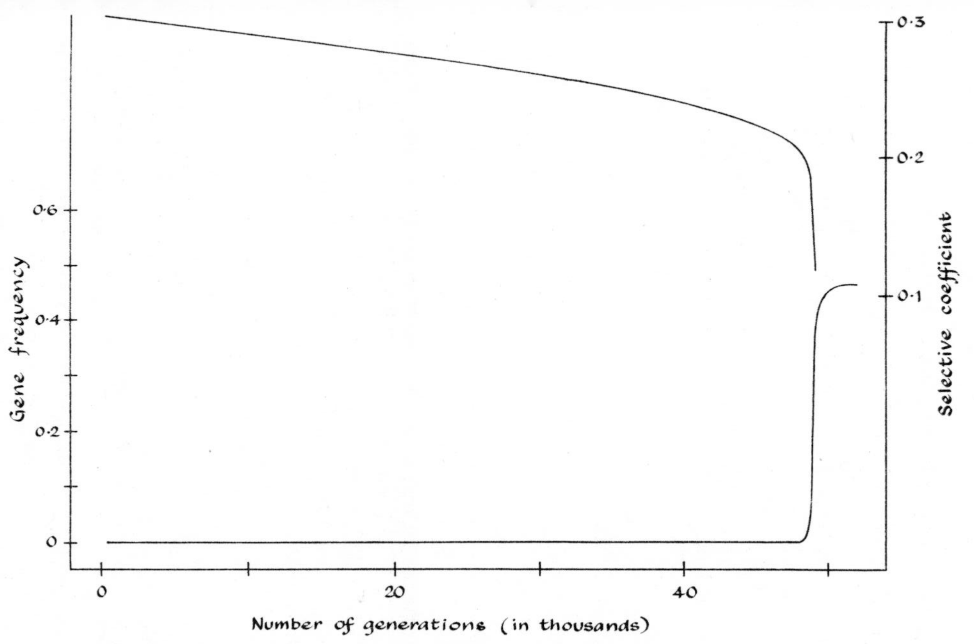

FIG. 8. The modification of the dominance of a deleterious mutant maintained by recurrent mutation. The upper curve shows the value of the selective coefficient of the heterozygote; the lower curve, the gene frequency of the mutant. The initial selective coefficients were: AA—0·2; Aa—0·3; aa—0·4. The initial additive variance was 0·005. (After O'Donald[141].)

shown in Fig. 8. Although O'Donald has suggested that variances of this magnitude and greater may be common in nature, we need to know much more about cases of this sort before generalizing from the model.

Increasing the Proportion of Heterozygotes

The principal limitation of all the foregoing models of dominance modification lies in the relative rarity of heterozygotes during the early stages of selection. This is a point emphasized particularly by Crosby.[32] If there are circumstances, however, in which the frequency of heterozygotes in the population is increased, then the selective development of dominance relationships will be facilitated.[63] This may occur during the spread of an advantageous gene in a population or as a result of balanced polymorphism.

The Spread of an Advantageous Gene

During the initial stages of the increase in frequency of a selectively favoured gene in a population, almost all of the individuals carrying the gene will be heterozygotes. Parsons and Bodmer[144] have calculated the ratio of heterozygotes to homozygotes at various stages of the process. Even as late as the stage when the advantageous gene has reached a frequency of 0·5, about five times as many heterozygotes as homozygotes will have been produced. This ratio appears to be rather insensitive to the difference in selective value between the favoured homozygote and the heterozygote. Bodmer and Parsons[6] therefore suggest that opportunities are enhanced for the evolution of the dominance of the advantageous gene.

As an example of the effectiveness of this mode of selection, Parsons and Bodmer[144] have cited Kettlewell's[93] investigations of the increased dominance of the gene for melanism in *Biston betularia*. Over the past 120 years the melanic (*carbonaria*) form of this moth has spread through the industrial parts of Great Britain and Europe. The change was particularly striking in the area around Manchester, where *carbonaria* increased in frequency from its status as a rare mutant in 1848 until it represented 98% of the population in 1895. Two lines of evidence indicate that during this spread the *carbonaria* gene evolved more complete dominance. First, Kettlewell[94] has examined 19th century specimens of *carbonaria* which are intermediate between modern melanics and the pale form. They have light lines and patches of white scales which are absent from the modern *carbonaria* of industrial areas today.

The second line of evidence comes from Kettlewell's[95] breeding experiments in which *carbonaria* was crossed with pale forms from populations in which the melanics are unknown. After three back-crosses to these stocks, the melanics began to show less extreme phenotypes, with white scales scattered over the wings. Thus it is clear that the selection by predators which has brought about the spread of the *carbonaria* gene has also enhanced its dominance over the pale form.

O'Donald[141] has considered the problem of the spread of an advantageous gene from the point of view of genetic variance mentioned above. Following the reasoning of Parsons and Bodmer he has considered the conditions under which dominance or over-dominance might arise. Again the outcome of selection depends on the magnitude of the additive genetic variance of the selective coefficient of the heterozygote. Dominance (or over-dominance, if an appropriate new optimal phenotype is possible) will arise provided that the genetic standard deviation in the fitness of the heterozygote is about 0·7 times the difference in fitness between the heterozygote and the homozygote. With lower values only semi-dominance will evolve.

O'Donald's calculations therefore would seem to set limits to the effectiveness of selection in modifying dominance during the spread of a favourable gene. With a variance of this magnitude about 15% of the heterozygotes already will be effectively dominant. Hence the outcome could be said to be inherent in the initial conditions. The real question is then whether or not such large variances actually exist in natural populations.

Polymorphism

The earliest exchange of views on the theory of dominance modification[63,76] emphasized the part that might be played by polymorphism. Since the existence of genetic polymorphism entails the production of large numbers of heterozygotes, the principal objection to the theory does not obtain in this case. Strictly speaking, the spread of an advantageous gene is an example of increasing the frequency of heterozygotes by means of polymorphism, in this case transient genetic polymorphism. But it is stable polymorphism which maintains a high frequency of heterozygotes, extending indefinitely the time during which selection may operate.

Many polymorphisms are of extremely long duration. There is good evidence, for example, that the polymorphism for shell pattern in the land snail *Cepaea nemoralis* has existed at least since the Pleistocene period.[40] Indeed, the extent of the parallelism between the genetic systems of *Cepaea nemoralis* and *C. hortensis* indicates that the polymorphism antedates speciation in the genus.[29]

Whether or not the modification of dominance is to be expected from the action of natural selection on a polymorphic species depends on the number of optimal phenotypes in the polymorphism and on the mechanism whereby the polymorphism is maintained. Sheppard[168] has pointed out that if a pair of alleles produce well adapted homozygotes superior to the heterozygote, the resulting disruptive selection would be expected to favour the evolution of dominance of one or the other of the alleles. As I have argued above, this outcome would be contingent upon the existence of some mechanism for maintaining the polymorphism, since disruptive selection alone will not usually provide a stable system.

A thorough analysis of the dominance relationships in a complex polymorphism has been provided by Clarke and Sheppard[25] in their work on the African swallowtail *Papilio dardanus*. This widespread butterfly occurs over much of the African continent with isolated races in Madagascar and Abyssinia. The males are monomorphic, but in most populations the females display an extravagant variety of mimetic forms resembling quite diverse species of models. The mimetic patterns are controlled by a long series of alleles at a single locus. With a few exceptions, all naturally occurring heterozygotes show complete dominance of one of the alleles involved.

In their extensive breeding of the various forms and races of *Papilio dardanus*, Clarke and Sheppard have clearly shown the role of the genetic background in determining the degree of dominance of a particular form. The most informative crosses are those between mimetic forms of the central African stock and the two isolated races. The race *meriones* from Madagascar is unique in that all the females are non-mimetic, having a black and yellow pattern similar to that of the males. Clarke and Sheppard have produced heterozygous females combining the *meriones* yellow with each of four continental mimics. In each case the hybrids are intermediate between the homozygotes, indicating that the dominance which is characteristic of sympatric forms is absent.

That this lack of dominance is not just the result of hybridity itself is shown by crosses of the central African mimics to the race *antinorii* from Abyssinia. Females of *antinorii* consist largely of the yellow male-like type, but a small proportion of mimetic forms occurs in most populations. The results of hybridization with *antinorii* male-like females differ strikingly from those with *meriones*. Provided that the mimetic form from central Africa is one which also occurs in the isolated Abyssinian race, then dominance relationships are quite regular. If, however, forms not found in the Abyssinian race are employed, then dominance tends to be incomplete.

These results indicate that dominance in *Papilio dardanus* has been evolved through the agency of disruptive selection. When alleles are brought together for the first time, they produce intermediate heterozygotes; but where they have a history of association, the heterozygotes usually produce the pattern of one or the other homozygote. Exceptions to the rule are found to involve combinations that are exceedingly rare in the wild, that give a third mimetic pattern as heterozygotes, or that occur in areas where models are scarce and the mimicry imperfect.[25]

A final point of interest that derives from Clarke and Sheppard's race crosses concerns the nature of the 'modifiers' responsible for the dominance of 'major genes'. In *P. dardanus* the non-mimetic forms are tailed as are other swallowtail butterflies. The central African mimics, however, carry a gene for absence of tails which is autosomal but limited in expression to the female sex. This gene is absent from the Madagascan and Abyssinian races, so that the mimetic forms in the latter are tailed. That there is selection for reduction of tail length in the Abyssinian mimics is suggested by a comparison of the relative lengths of tails in different forms in the two races (table 8).

TABLE 8

Lengths, in mm, of tails of Papilio dardanus *from Madagascar and Abyssinia. The Madagascan race has no mimetic forms. (After Clarke and Sheppard*[25]*.)*

	Madagascar	Abyssinia
Males	14·4	14·2
Non-mimetic females	14·8	12·5
Mimetic females	—	9·0

In race crosses involving the gene for taillessness, there is again a difference between the Madagascan and the Abyssinian hybrids with the central stock. All the Madagascan heterozygotes have short tails while about half of the Abyssinian ones have no tails at all. The backcross of the Abyssinian hybrids to the Abyssinian stock also produces heterozygotes about 50% of which are tailless. It can be concluded, therefore, that selection for taillessness in the Abyssinian mimics has produced a genotype with 'modifiers' capable of enhancing the dominance of the 'major gene' at its first occurrence. In such cases where polymorphism develops to supercede a previously inefficient genetic mechanism, the evolution of dominance may be nearly complete when the polymorphism first arises. Modifiers, therefore, would appear to be nothing more than the ordinary genes with small effects that contribute to the genetic variance of quantitative characters.

The Problem of Rare Dominants

One final aspect of dominance modification is worth consideration since it involves an apparent exception to Fisher's theory. Haldane[77] has called attention to the fact that rare genes may be dominant to their more common alleles. He cited *Primula sinensis* in which 8 out of 34 rare alleles are dominant or semi-dominant, and *Papaver rhoeas* with 2 out of 8. Since these species are normally outbred, Haldane argued that they should therefore have had ample opportunity for selection to modify the heterozygotes. At first glance, these cases would seem to contradict the expectation from Fisher's theory, *i.e.* that the most favourable allele, and hence the commonest, should be the one to evolve dominance.

A wide variety of polymorphisms show the phenomenon of a 'universal recessive' with several rarer dominants. In the land snail, *Cepaea nemoralis*, the common yellow banded form is recessive to pink and brown at one locus, to unbanded at what is presumably a separate but closely linked locus, and to midbanded at a third.[15] In another snail, *Partula taeniata*, the common pale unbanded form is recessive to brownish-purple at one locus and to banded at another.[138] The marine isopod, *Sphaeroma rugicauda*, shows a whole series of rare forms dominant to the common *grey*.[198]

Sheppard[168] has argued that there are at least two possibilities for reconciling these observations with Fisher's theory. In the first case he has considered a polymorphism resulting from heterozygous advantage at a physiological level, with the morphological expression of one homozygote conferring a disadvantage and the heterozygote intermediate. The disadvantageous morphological characters of both would tend to be selectively modified in the direction of the superior homozygote, but the heterozygous advantage would prevent the modification of the heterozygote. The result would be for the morphologically disadvantageous homozygote to approach the heterozygote in appearance.

The principal objection to such a scheme, as Clarke[20] has pointed out, is the assumed interdependence of the deleterious morphological character and the stabilizing heterosis. Sheppard himself has argued repeatedly that dominance is a property of characters and not alleles, and therefore there would seem to be no reason why in this case the unfavourable character should not become completely recessive without disturbing the heterozygote advantage.

Sheppard's second explanation embraces the situation in which there are three separate phenotypes, only two of which are optimal. The evolution of dominance in such a case will minimize the loss of fitness by the sub-optimal type. Sheppard argues that the most recently occurring allele will be the one to attain dominance. Since it must be selectively favoured in the heterozygous state in order to become established at all, the sub-optimal form must be the new homozygote. The new allele would therefore become dominant by the modification of the homozygote to resemble the heterozygote.

Both of these mechanisms imply the evolution of dominance in the opposite direction from the usual one, *i.e.* modification of the homozygote rather than the heterozygote. This is not a serious objection since recent work by Ohh and Sheldon[211] has shown that, at least in the case of *Hairy-wing* in *Drosophila melanogaster*, artificial selection can produce dominance by altering either the mutant homozygote or the heterozygote. Sheppard's models could, therefore, account for instances of rare dominants, but in neither case is there a direct relationship between the rarity of the allele and its dominance.

Clarke[20] has suggested an ingenious scheme which is compatible with Fisher's theory and which nevertheless predicts the occurrence of polymorphisms with dominance of rare alleles. His model, based on Batesian mimicry, combines both frequency-dependent and independent selection. Clarke points out that in any polymorphism there will be a set of phenotypic frequencies, which he terms 'foci', at which the frequency-dependent components of the selective values of the types are equal. If the focal frequency of a distinct, intermediate heterozygote is low, then there will be selection for the dominance of whichever allele is rarer in the population. Thus the existence of a frequency-independent component of selection, such as reduced viability of one homozygote, would result in selection for the dominance of that allele. The dominance would be established only with respect to the character responsible for the frequency-dependent selection however. It would not, of course, extend to the frequency-independent component. Indeed dominance modification should ensure the recessiveness of the reduced viability. It is perhaps worth remarking that in such an event the outcome would be the equivalent of heterozygous advantage since the heterozygote would combine the favourable characters of both homozygotes, *i.e.* optimal viability and rarity.

The experimental investigation of dominance modification in a system characterized by frequency-dependent selection remains to be undertaken. Since it will involve the simultaneous manipulation of two relatively poorly understood phenomena, the technical problems are likely to be severe.

Conclusions

From the evidence which has accumulated since Fisher's theory was propounded in 1928, it is clear that the experimental modification of dominance can easily be achieved. It seems equally clear that under optimal conditions modification will take place in natural populations. The chief limitations on the process are the availability of genetic variance, the occurrence of sufficient numbers of heterozygotes, and time. The development of dominance will be facilitated whenever circumstances materially increase the frequency of heterozygotes in the population, as in the case of transient or stable genetic polymorphism.

It is probable that in many cases dominance is inherent rather than evolved. A primary defect in a structural gene would be expected to result in immediate recessiveness, although the existence of suppressors and regulator genes suggests that even at this level selective modification of dominance is possible.

The occurrence of rare dominants, apparently contradictory to the idea of evolved dominance, may be explained if morph frequencies are determined in part by frequency-dependent selection.

5: The Wallace Effect

> 'The simplest case to consider, will be that in which two forms or varieties of a species, occupying an extensive area, are in the process of adapting to somewhat different modes of life....
>
> 'Now, let us suppose that a partial sterility of the hybrids between the two forms arises, in correlation with the different modes of life and the slight external and internal peculiarities that exist between them, both of which we have seen to be real causes of infertility. The result will be that, even if the hybrids between the two forms are still freely produced, these hybrids will not themselves increase so rapidly as the two pure forms; and as the latter are, by the terms of the problem, better suited to their conditions of life than are the hybrids between them, they will not only increase more rapidly, but will also tend to supplant the hybrids altogether wherever the struggle for existence becomes exceptionally severe.' A. R. Wallace: *Darwinism.*[191]

The arguments of Chapter 3 suggested that there are two possible avenues of escape from the dilemma posed by selection against heterozygotes. One of these solutions, the evolution of dominance, has been considered in some detail. The other possibility is the division of the population into two isolated portions which no longer interbreed. A particularly interesting and controversial question at the present time is whether such a separation is possible without prior spatial isolation. A subsequent chapter will be devoted to this problem. However, a more limited question must be considered first, since the possibility of sympatric speciation is in large measure contingent on the answer: can the process of natural selection lead to the origin and perfection of isolating mechanisms?

The distinction between the principal alternative theories of the origin of reproductive isolation was drawn quite early in the debate on natural selection. Although he was primarily interested in the transformation rather than in the division of species, Darwin[38] considered that isolation came about simply as a by-product of evolutionary change. On the other hand Wallace[191] contended that as soon as two populations had diverged to the extent that hybrids between them were less well adapted than either parent, then natural selection would tend to eliminate the hybrids. Such a process would serve to protect the evolving adaptation of the incipient species. Grant[75] has proposed, therefore, that this selection for reproductive isolation be termed the *Wallace Effect*.

There is, of course, nothing in the two theories that makes them mutually exclusive. Indeed there can hardly be disagreement on the occurrence of reproductive isolation as a by-product of divergence in isolation. The question to be considered, then, is whether the Wallace Effect can be detected in specific cases as an additional element in the process of speciation and if so, whether it occurs sufficiently frequently to give it special significance.

There are at least four types of evidence that are relevant to the Wallace Effect:[75] (1) Artificial selection for reproductive isolation. (2) Changes in the frequency of hybridization within hybrid zones. (3) Comparison of the strength of isolation between two species which are sympatric in some places and allopatric in others. (4) Comparisons between species within a group having some sympatric species and some allopatric species. Following a brief account of some findings on the development of isolation without selection, each of these phenomena will be discussed in turn.

Incipient Isolation

There is little need to stress the possibility that reproductive isolation may develop between spatially separated populations. Such a process is implicit in the formation of any group of species which form what Mayr[124,125] has called a superspecies. Two other kinds of evidence, however, help to make it explicit.

In the first instance, it is often possible to show that a widespread species consists of populations which show increasing reproductive isolation with increasing distance from each other. In the *Drosophila funebris* group, for example, Patterson and Stone[146] have described a chain of populations beginning with *D. subfunebris* in California and extending through three subspecies of *D. macrospina* to the eastern United States. Individuals from adjacent populations may be crossed without difficulty, but fertility between populations decreases with increasing distance. Rick[160] has found a similar situation in South American tomatoes. *Lycopersicon peruvianum*, a widespread species in Peru and northern Chile, is represented in the northern portion of its range by the variety *humifusum*, which is morphologically distinct and is not crossable with the majority of the more southerly populations.

There are populations, however, which are intermediate in morphology and which may be crossed to both northern and southern forms. The degree of isolation of populations is proportional to the distance between them.

The second line of evidence consists of laboratory data on incipient reproductive isolation between stocks of animals which have been kept separate for varying periods of time. Koref-Santibañez and Waddington[105] tested a number of stocks of *Drosophila melanogaster* of three different kinds for incipient isolation. Some of the lines had been selected previously for either high or low chaeta number. Others were inbred strains which had been brother/sister mated for 57 generations. Some of these had been irradiated before inbreeding. Still others were isogenic stocks, some inbred and some raised in mass cultures.

Stocks were tested by means of Dobzhansky's 'male choice' method in which male flies are placed in mating vials with females of their own and of some other type. The initiation and duration of the first copulation are recorded, after which the females are checked for the presence of sperm. A smaller number of 'female choice' trials were also run.

The results varied considerably from group to group. Among the selected stocks mating took place essentially at random. Small significant departures from equality of homogametic and heterogametic matings occurred in several cases, but almost equally frequently in each direction. With the inbred lines, there are significant homogametic tendencies in each of the three unirradiated lines. The results from the isogenic stocks are complicated by the presence of the mutant *yellow* which apparently makes a male almost completely unacceptable to non-yellow females, although yellow females have enhanced receptivity.

It is difficult to give a general interpretation of these experiments other than to say that they indicate the possibility of the initiation of incipient isolation in a relatively short span of time. There is also the further suggestion that inbreeding may be more important in this respect than differential selection.

These observations have been extended by Hoenigsberg and Koref-Santibañez[86] to other outbred and inbred strains of *D. melanogaster*, the latter resulting from 404 generations of brother/sister mating. Although Merrell[129] has pointed out that some of the apparent effects are attributable simply to the less frequent mating of the inbred females and to the longer period of courtship which they require, there is nevertheless a residual core of sexual isolation among strains.

More recently, Ehrman[50] has described the development of weak but significant isolation in laboratory populations of *D. pseudoobscura*. The populations were established by Vetukhiv from a synthetic stock combining more than 40 strains from four localities. Six successive groups of offspring from the same parents were the founders of six populations which were then maintained in isolation for four and a half years. Two populations were kept at each of three temperatures, 16°, 25° and 27°C.

After Vetukhiv's death, Ehrman undertook the analysis of the resulting stocks. Subcultures of each stock were reared under standard, optimal conditions so that the developmental environments of test animals should be strictly comparable. 'Male choice' experiments were used exclusively, ten males being paired with ten similar females and ten from a different stock. In all, twelve different choices were analysed at each of two different test temperatures. Of the 24 tests, nine showed a significant excess of homogametic matings. Perhaps the most striking result was that seven of the nine significant results occurred between pairs of populations reared at the same temperature. Thus there is again no suggestion that natural selection for the different environmental temperatures played a role in the divergence.

In addition to this kind of gradual build-up of isolation in laboratory stocks, there are other examples in which the changes are rather sudden and dramatic. Dobzhansky and Pavlovsky[46] describe such a case in *Drosophila paulistorum*. This species in its natural state consists of a group of incipient species which may overlap without interbreeding but which are at least potentially capable of exchanging genes by way of transitional races or strains. The laboratory strain in which the sudden change occurred, Llanos, is one which was originally classified, by means of hybridization tests, as a member of the Orinocan incipient species. When retested after four years in culture, the 'new' Llanos strain was found to have acquired reproductive isolation from the Orinocan stocks, all hybrid males being sterile. The separation has apparently been brought about by the infection of Llanos with a symbiotic microorganism. Although it produces no pathological effects within the Llanos strain, the symbiont may cause sterility or death if it is transferred to other strains by injection. Interestingly enough the new Llanos strain still mates readily with the Orinocan stocks, although a strain from the wild which is interfertile with new Llanos is ethologically isolated from Orinoco. Dobzhansky and Pavlovsky suggest, therefore, that the sterility represents the initial isolating mechanism and that the ethological isolation is secondarily acquired.

All of these studies above share one common denominator. In no case is there any likelihood of direct selection for isolation. Hence the conclusion is inescapable that in many cases, separation alone, particularly if extended over a protracted period, is sufficient to produce reproductive barriers.

Artificial Selection for Isolation

Incipient reproductive isolation in spatially separated populations implies the existence of genetic variability in isolating mechanisms. The question which naturally arises is whether selection is capable of making use of such variability to enhance isolation. If so, then wherever crosses tend to produce offspring of lowered fitness, the Wallace Effect should occur.

Direct evidence on this point comes from experiments on artificial selection in laboratory populations. A pioneer study of this kind has been carried out by Koopman[104] on a mixed population of *Drosophila pseudoobscura* and *persimilis*. These two species cross fairly readily in the laboratory even though the hybrid males are sterile and the hybrid females give rise to offspring of drastically lowered viability. In order to distinguish the parental species from the hybrids, Koopman found it necessary to mark them genetically, *D. pseudoobscura* with the mutant

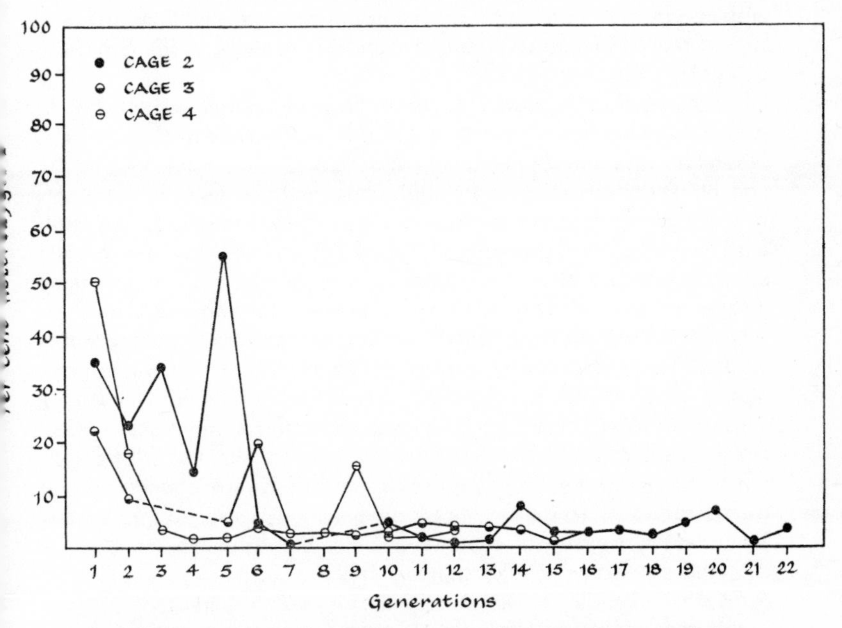

FIG. 9. The progress of artificial selection for reproductive isolation between *Drosophila pseudoobscura* and *D. persimilis*. The decrease in the proportion of hybrids is shown for three separate laboratory populations. (After Koopman[104].)

glass and *D. persimilis* with *orange*. Thus any wild-type offspring in the cultures were hybrids.

In each generation a mixture of virgin males and females of each species (usually 200 to 800 flies, equally distributed by species and sex) were allowed to mate in mass culture. Any hybrids which appeared among the offspring were discarded before the next generation was set up. This regime confers effective lethality on hybrids and therefore represents very heavy selection against hybridization. It is perhaps not surprising that the proportion of hybrids among the offspring decreased rapidly in a very few generations. Fig. 9 shows the progress of selection. In all cases the proportions of wild-type flies had fallen to less than 5% by the sixth generation.

The experiments on isolation between *D. pseudoobscura* and *persimilis* have been extended by Kessler[92] who has selected directly for behavioural characteristics leading to both increased and decreased isolation. By choosing as parents those animals which gave rapid interspecific matings for a 'low isolation' line and those which did not mate interspecifically at all for a 'high isolation' line, Kessler achieved a response in both directions. The chief limitation, in the 'low isolation' line, appeared to be the continued reluctance of the *persimilis* females to accept *pseudoobscura* males.

The speed with which increased isolation develops between *D. pseudoobscura* and *persimilis* stands in marked contrast to the results of laboratory experiments involving two strains of one species. Wallace[192] and Knight, Robertson, and Waddington[101] have described attempts to produce reproductive isolation within *D. melanogaster*. In the latter experiment stocks of *ebony* and *vestigial* flies were extracted from a strain in which both loci were segregating. Virgin *ebony* and *vestigial* flies were mixed to provide the parents of each generation. Any wild-type flies, resulting from crosses of the two parental types, were discarded. The proportion of wild-type flies decreased, but much more slowly than in the interspecific cross. The results are shown in Fig. 10. Between the 20th and 30th generations the females were tested to see which type of male each individual accepted. The results are shown in table 9. It is interesting that the controls show a slight (non-significant) tendency toward homogamy. This effect is greatly intensified as a result of selection ($P \ll 0{\cdot}001$), although nothing like complete reproductive isolation was ever achieved.

A final example of increasing isolation in *Drosophila* is interesting in that it combines the initiation of isolation without selection, followed by a further enhancement as a result of a single selective event. Henslee[84] has described a parthenogenetic strain of *Drosophila mercatorum* derived from a cross of stocks from New York and El

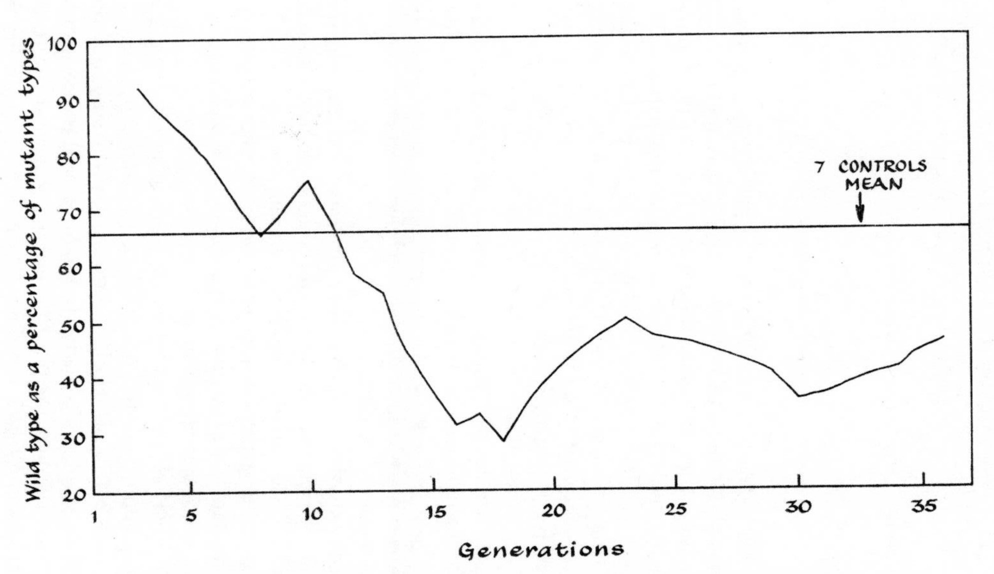

FIG. 10. The progress of artificial selection for reproductive isolation between two strains of *Drosophila melanogaster* marked with the genes *ebony* and *vestigial*. The number of double heterozygotes (wild type) is expressed as a percentage of the sum of *ebony* and *vestigial* flies emerging. (After Knight, Robertson, and Waddington[101].)

TABLE 9

Matings between ebony (e) *and* vestigial (vg) *mutants of* Drosophila melanogaster *after* 20 *generations of selection for reproductive isolation. (After Knight, Robertson, and Waddington*[101]*.)*

	Females	Inseminated by	
		e	vg
Controls	e	71	69
	vg	41	63
Selected stocks	e	151	108
	vg	77	142

Salvador. The strain was selected for increased parthenogenetic reproduction for three generations and was then maintained in mass culture without males for 52 generations. Mating tests of the 'male choice' type were carried out with ten males of one of the parent strains confined with ten females of each of two types to be tested. Under these conditions, females of either of the original stocks were inseminated much more frequently ($P < 0{\cdot}01$) by either type of male than were females from the parthenogenetic strain.

In a further experiment, two lines derived from the parthenogenetic strain were tested against the parthenogenetic strain itself. The lines were set up from single females which remained unmated in a choice experiment like the one above. One of these lines was not detectably different, but the other was inseminated much less frequently than the parthenogenetic strain ($P \leqslant 0{\cdot}001$).

Quite apart from their significance with respect to the origin of parthenogenesis, the experiments show how isolation and selection can work synergistically to initiate incipient reproductive isolation. The principal effect of the parthenogenesis is probably to relax selection for mating efficiency and compatibility in the stocks, in which mating is not required for successful reproduction.

Some indication of the opportunism by which selection may operate to bring about isolation can be seen from an experiment on maize carried out by Paterniani.[145] Two strains carrying different marker genes were planted alternately in the same field for six generations. The kernels selected for planting in each generation were the products of intra-strain crosses and were chosen from ears showing the lowest proportion of inter-strain crossing. Initially the proportion of intercrossing was quite high, 35·8% and 46·7% for the two strains. Selection for isolation proceeded with remarkable rapidity. After six generations the proportions had dropped to 4·9% and 3·4% respectively. Table 10 shows the progressive nature of the change.

TABLE 10

Average per cent of intercrossing in maize of the White Flint and Yellow Sweet varieties selected for reproductive isolation. (*After Paterniani*[145].)

Generations of selection	No. of Plants	% Intercrossing
White Flint orig.	295	35·8
,, ,, I	400	24·9
,, ,, II	477	14·0
,, ,, III	489	10·3
,, ,, IV	605	9·2
,, ,, V	590	4·9
Yellow Sweet orig.	232	46·7
,, ,, I	253	30·6
,, ,, II	368	35·1
,, ,, III	350	9·3
,, ,, IV	389	10·6
,, ,, V	485	3·4

Perhaps the most significant aspect of this work is the identification of the principal mechanism responsible for the reduction in crossing. The initial populations of the two varieties hardly differed at all in flowering time, but the selected populations showed a difference in mean flowering times of more than seven days. In plants with two ears, the later ear of the earlier strain and the earlier ear of the later strain were much more subject to inter-strain crossing than were the ears more widely separated in time.

The limitation of all of these experiments on artificial selection for reproductive isolation, of course, is the question of their relevance to natural conditions. At least in the case of changes in flowering time, there is evidence, which will be discussed in the following chapter, that similar effects are important in wild populations. In any event, the experiments indicate that genetic resources exist for implementing the Wallace Effect should the external circumstances provide the opportunity.

Hybrid Zones

In natural populations, the most dramatic phenomenon bearing on the origin of reproductive isolation is the occurrence of old, stable hybrid zones. The evidence provided by such cases is difficult to reconcile with selection for reinforcement of isolating mechanisms. On the one hand, the failure of the two hybridizing groups to merge seems to indicate that recombination results in less successful genotypes, which should be removed by selection. On the other, the continued hybridization implies that the Wallace Effect is not operating efficiently.

The classic case in point is that of the European crows.[128] The Carrion Crow (*Corvus corone*) is entirely black and is found throughout most of western Europe. The Hooded Crow (*Corvus cornix*) is grey with black head, wings, and tail. It inhabits Scandinavia, Eastern Europe, and much of the Mediterranean coast. The two species contact each other in a line which crosses central Scotland, Denmark, Germany, and Austria, swinging around the southern side of the Alps to reach the Mediterranean at Genoa. All along this line there is a narrow zone of hybridization which averages between 75 and 150 km in width. Apparently the zone may shift without losing its narrow definition.[89]

There seems to be no evidence that the degree of hybridization is changing. Mating appears to be random in mixed populations, but the resulting recombinant types only appear very rarely outside the zone. Dobzhansky[41] has suggested that there has been a contraction of the hybrid zone within recorded history but there is little or no evidence to substantiate that idea. In fact the narrowness of the hybrid zone both in the north and in the south suggests that the relative age of the contact between the two species has little effect on the width. Therefore it is unlikely that the hybridization is either spreading or contracting.

The geographical relationships of the Carrion and Hooded Crows are almost certainly the result of the expansion of populations from glacial refuges at the end of the Pleistocene glaciation. Ecologically tied to areas with trees, the two crows would have been confined by the arctic conditions of central Europe to the southeastern and southwestern extremes. There they developed their distinctive characteristics which have been maintained despite subsequent expansion and hybridization.

If the crows represented a single, unusual case, then it might be possible to argue that special circumstances were responsible. On the contrary, hybrid zones are common where closely related species occupy adjacent territories. The documentation is impressive for birds, probably reflecting the state of our knowledge rather than the uniqueness of the group. Palearctic examples could be cited in nuthatches (*Sitta*), shrikes (*Lanius*), and wagtails (*Motacilla*), and American cases in grackles (*Quiscalus*), flickers (*Colaptes*), and towhees (*Pipilo*). Other groups might equally well be mentioned such as pocket gophers (*Thomomys*), sticklebacks (*Gasterosteus*), and snails (*Thais*). References to these and many other cases are given by Sibley[173], Mayr[125] and Remington.[159]

The principal question, which remains unanswered in most of these examples is whether or not any changes are taking place with the passage of time in these hybrid populations. Only very rarely is the historical record sufficiently detailed to detect a change in the frequency of hybridization. Vaurie[190] has one example in tits of western Russia. Between 1870 and 1900 the Azure Tit (*Parus cyanus*) from eastern Russia moved into an area occupied by the closely related Blue Tit (*Parus caeruleus*) producing a high proportion of hybrids in the region of overlap. Over the past sixty years, however, fewer and fewer hybrids have occurred. A similar situation has been reported by Bauer[3] in the woodpeckers *Dendrocopus major* and *D. syriacus* in southeastern Europe and Asia minor.

Unfortunately, not even the direct observation of a reduction in the frequency of hybridization is sufficient to demonstrate selection for reproductive isolation. Bauer[3] has offered an alternative explanation based solely on the numerical relationships of the two species of woodpeckers. He points out that during its expansion, the Syrian Woodpecker would be greatly outnumbered along the advancing front. The probability of its finding a conspecific mate would therefore be reduced and hybridization would follow. With increasing numbers, opportunities for appropriate matings would increase, with a consequent reduction in hybridization.

It appears, therefore, that the evidence for the selective increase of reproductive isolation within hybrid zones is either negative or equivocal. Since short-term observations are unlikely to produce significant results, it would seem to be important to initiate in selected hybrid zones careful studies in which periodic sampling would extend over a sufficiently long time to pick up small but consistent changes in the frequency of hybridization.

There is another aspect of the general picture of hybridization, however, which provides indirect evidence on the selective reinforcement of isolating mechanisms. Sibley[172] has called attention to several groups of birds which share certain characteristics with no immediately obvious functional relationship. The associated characters are (1) strongly marked sexual dimorphism with highly specialized signal characters, (2) polygamy and/or transient pair bonds, (3) sympatry with related species, and (4) occasional hybridization with these related species. All or most of these traits are found among members of the birds of paradise, the hummingbirds, the pheasants, the grouse, the manakins, and the ducks of the genus *Anas*.

Sibley has argued that sexual selection and the reinforcement of isolating mechanisms are responsible for this constellation of characters. Competition among males, enhanced by polygamy, results in the development of sexual dimorphism. The presence of closely related species with which hybridization is possible should then bring about divergence among the special signal characters which, of course, would serve as isolating mechanisms. This hypothesis leads to the prediction that species which are geographically isolated from their close relatives should show a reduction in sexual dimorphism. Isolated species of the genus *Anas* provide notable examples. The mallards of Hawaii and Laysan Island and the pintails of Kerguelen Island and Crozet Island have lost their distinctive male plumages. In the absence of continuing selection for the perfection of isolating mechanisms, counterselection has provided the males with the cryptic pattern found elsewhere only in the females.

Partially Sympatric Species

Most of the evidence bearing upon the Wallace Effect in natural populations is indirect. It consists of observations of the relative strength of isolating mechanisms in sympatric and allopatric populations of two or more species. Grant[75] has made a useful distinction between two groups of phenomena of this kind. The first employs pairs of species which have a partially overlapping distribution. A comparison is made of the ease of crossing of individuals of the two species from areas where each occurs alone and from areas where the two are sympatric. The second consists of similar comparisons of several pairs of species from a closely related group, some pairs of which are sympatric, and other pairs, allopatric. The first case will be discussed in this section, and the second in the following one.

A pioneering study of the relative strengths of isolating mechanisms in sympatric and allopatric populations was carried out by Dobzhansky and Koller[45] on the western American species *Drosophila pseudoobscura* and *D. miranda*. They found that *D. miranda* from the Puget Sound region showed greater sexual isolation from sympatric *D. pseudoobscura* than from allopatric populations, although the interpretation of these tests is complicated by the existence in California of sympatric *miranda* and *pseudoobscura* with rather less isolation.

Grant has questioned the significance of results of this kind, pointing out that only a comparative study of many cases within a group of species can be meaningful since individual observations may go either way. He cites the findings of Patterson, Stone, and Griffen[146] on the *virilis* group of *Drosophila* in which *americana* and *texana* cross easily with *virilis* from China but poorly with *virilis* from Victoria, Texas. This apparent parallel to the *pseudoobscura* case becomes ambiguous since *virilis* from Galveston, Texas, crosses just as easily as *virilis* from China. Grant also cites similar contradictory evidence in the plant genus *Gilia*. *G. modocensis* and *G. sinuata* are more difficult to cross when sympatric than when allopatric, but exactly the reverse is true of *G. splendens* and *G. australis*.

The experiments of Dobzhansky and his co-workers[44,51] on the group of six incipient species in the *Drosophila paulistorum* complex provide at least a partial answer to Grant's criticism. Five of the six yield a total of eight comparisons between allopatric and sympatric populations of pairs of semispecies. Ehrman[51] has studied all of these combinations by direct observation. Essentially the method is to enclose in a mating chamber 10 males and 10 virgin females of each of the two species to be compared. Each mating is scored, and the observations are continued until about half of the females have been mated. The flies are then discarded and a new batch introduced. For each comparison an index of isolation may be calculated which is simply the proportion of homogametic matings minus the proportion of heterogametic matings. Thus it varies from +1 indicating complete isolation, through 0 for random mating, to −1 for complete dis-assortative mating.

Ehrman found that in seven of the eight *paulistorum* comparisons, the index of isolation was higher for the sympatric than for the allopatric populations. In the eighth the index was greater for the allopatric populations, but not significantly so. Overall the average index of the sympatric populations was 0·85 compared with 0·67 for the allopatric ones. These figures demonstrate convincingly that at least for this species complex, sympatry is associated with increased sexual isolation.

Another group which provides enough comparisons for an overall evaluation of the effects of sympatry is the frogs and toads. For several reasons these are favourable animals for tests of the Wallace Effect. In the first place it is rather simple to carry out artificial crosses in order to estimate the viability and fertility of the resulting offspring. In the second, reproductive isolation between sympatric species depends heavily on the differentiation of mating calls. This provides a very convenient means of analysis, since the technique of sound spectrography yields quantitative data on call duration, rate of repetition, pulse rate, and dominant frequency. Combining these two sorts of data it is possible to compare the relative strengths of pre-mating and post-mating isolating mechanisms in various combinations of sympatric and allopatric populations.

It is not surprising, therefore, that a great deal of work has been carried out on isolating mechanisms in anurans both in America and in Australia. The core of the evidence for the American species has been summarized in a review by Blair.[4]

Perhaps the most interesting overall result is the number of examples in which differentiation of ecology and mating call precedes the appearance of marked hybrid sterility or inviability. In the *Bufo americanus* group, for example, there are six species with quite distinct calls (*B. houstonensis* is an exception). As Fig. 11 shows, there is little or no immediate post-mating isolation beyond a suggestion of slightly reduced viability in the F_2 generations of some crosses. Much the same is true of the species which have been tested from the *B. valliceps* group.

SYMBOLS

M+ = Metamorphosis (not raised)

A+ = Adult (not tested)

F = Fertile

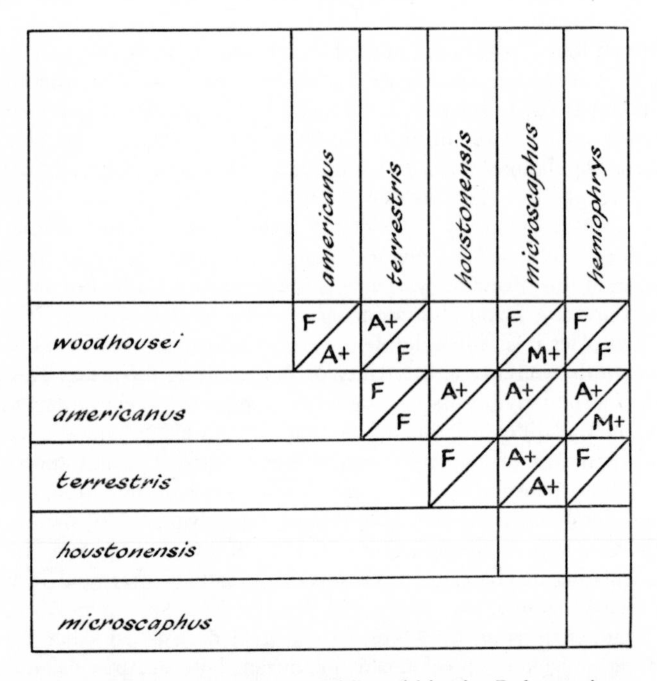

FIG. 11. A summary of relationships within the *Bufo americanus* species group as shown by interspecific crosses. The upper left symbol of each box indicates the cross with the species on the left as the female and the lower right symbol, with that species as the male. (After Blair[4].)

In the family Hylidae, a number of species pairs show this phenomenon as well. *Pseudacris triseriata* and *P. clarki*, which overlap in eastern Texas, show striking differences in call and in ecology, yet interspecific hybrids are fertile.[130] *Hyla cinerea* and *H. gratiosa* and also *Hyla chrysoscelis* and *H. avivoca* provide two more examples. In each case the range of the first species exceeds that of the second, and hybrids are fertile, but there is ecological isolation and call differentiation. In an artificial environment created by the building of fish ponds near Auburn, Alabama, the isolation between *H. cinerea* and *H. gratiosa* has broken down.[127]

Against this general background of the pre-eminence of premating isolation must be placed a large number of cases in which call differentiation is greater in areas of sympatry than in allopatry. Blair[4] has catalogued examples in *Microhyla*, *Acris*, *Pseudacris*, *Bufo*, and *Scaphiopus* in which calls become more distinct in areas of overlap.

With two Australian species of the genus *Hyla*, Littlejohn and Loftus-Hills[117] have recently shown that the behavioural responses of the animals follow closely the characteristics of the call. *Hyla ewingi* and *H. verreauxi* are two closely related species in southeastern Australia. The principal ranges of the two are allopatric, but across the border of Victoria and New South Wales each species sends a long tongue into the territory of the other. In the areas of allopatry the mating calls are extremely similar, but the deeper each tongue penetrates into the territory of the other species, the more distinct the calls become, particularly in pulses per note and in pulse repetition frequency. These characteristics are clearly shown by the oscillograms in Fig. 12.[116]

In order to demonstrate the functional significance of these changes, Littlejohn and Loftus-Hills have scored the responses of females in an apparatus providing a choice between calls of two different types. All the females came from the western part of the area of overlap (Fig. 12b) where the call of *H. ewingi* scarcely differs from the neighbouring allopatric *H. ewingi* populations but where that of *H. verreauxi* attains its maximum divergence. The results, though few in number, are clear-cut. *H. ewingi* females chose sympatric *ewingi* calls over sympatric *verreauxi* without exception, but failed to discriminate between sympatric *ewingi* and allopatric *verreauxi*. *H. verreauxi* females unerringly chose *verreauxi* over *ewingi* from sympatric populations, and also chose sympatric *verreauxi* over allopatric *verreauxi* in every case. Although the clinching case in which allopatric *verreauxi* and sympatric *ewingi* from the *eastern end of the overlap* are offered to western sympatric *ewingi* was not investigated, it can hardly be doubted that *ewingi* would choose the inappropriate species.

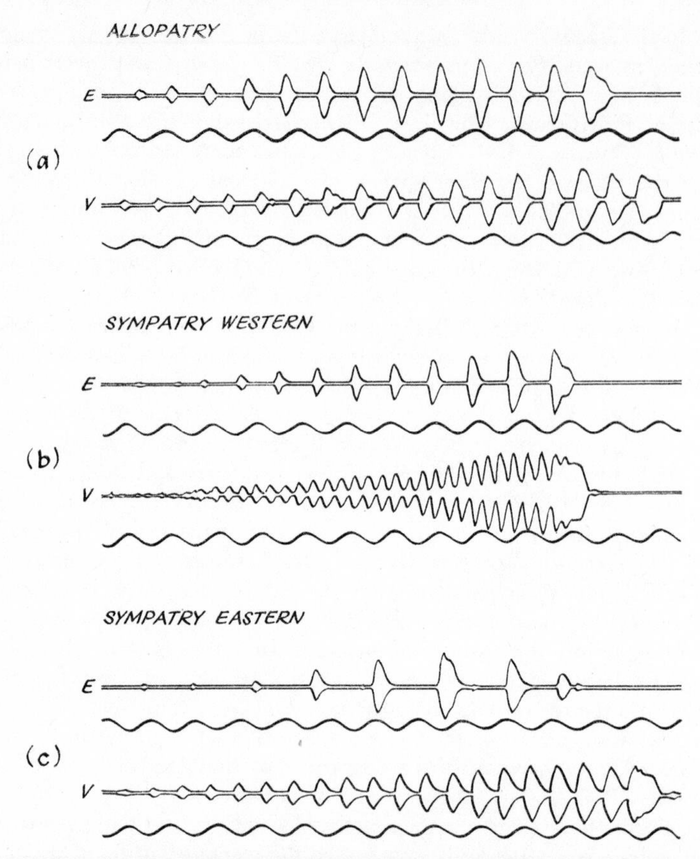

FIG. 12. Oscillograms of mating calls of *Hyla ewingi* (E) and *H. verreauxi* (V) from areas of allopatry and sympatry. The lower trace in each oscillogram is a time base of 50 cycles per second. (After Littlejohn[116].) [N.B. Patterns represent outlines of tracings.]

The work with the *ewingi* complex has brought to light one very puzzling finding. Watson and Martin[197] report that hybridization between the two species is less successful between sympatric populations than between allopatric ones or between an allopatric and a sympatric population. The data leave something to be desired, since the controls for one of the sympatric crosses also had a high proportion of abnormal embryos, and survival to metamorphosis was very low in all cases; but the suggestion does raise the question of possible selection for post-mating isolating mechanisms.

Although it must be conceded that in general selection will only favour isolating mechanisms which operate in the parental generations,[74,136] there is at least one case in which selection for a post-mating mechanism is suspected. Stephens[180] has described the genetics of the *corky* syndrome in the new world cottons, *Gossypium hirsutum* and *barbadense*. This condition, characterized by bushy, stunted plants with corky stems, results from joint action of the alleles ck^x and ck^y, one derived from each species. These alleles are found primarily in the West Indies where the two species have a history of joint cultivation, but they are largely absent from allopatric populations.

Stephens suggests that the unusual conditions surrounding the joint cultivations of these perennial stocks have resulted in selection for the corky alleles, which operate as a post-mating isolating mechanism. *G. hirsutum* and *G. barbadense* normally form heterotic hybrids, but subsequent F_2 and backcross generations are markedly inferior to both parents and F_1 hybrids. Hence selection of seeds from the most vigorous plants inevitably leads to deterioration of the stock through hybrid breakdown. In stocks with the *corky* alleles, however, F_1 hybrids are obviously inferior and would not be chosen as parents.

The advantage to the grower of a stock containing fixed *corky* alleles is obvious. The difficulty lies in establishing the alleles in the population. At low frequencies they are at best neutral or they would be expected to occur in all populations. Hence we are faced with the problem of the initial, and joint, increase in the frequencies of *corky* in the two species, a problem which remains unsolved.

A final example of divergence in sympatry is parallel in every way to the case of *Hyla ewingi*, only this time in plants. Levin and Kerster[109] have investigated the rather curious distribution of flower colours in *Phlox pilosa*. When growing alone *P. pilosa* normally bears pink flowers, but in mixed populations with *P. glaberrima* most of the flowers are white. To demonstrate that the altered colour is in fact a 'behavioural' isolating mechanism operating via pollinating insects, Levin and Kerster have transplanted enough pink *pilosa* into a predominantly white population to raise the frequency of pinks to 25%. Under these conditions pollen from *P. glaberrima* was deposited on 30% of the stigmata of pink *P. pilosa* but on only 12% of the stigmata of white flowers. The pink *pilosa* flowers received 4·8 times as much *glaberrima* pollen as white *pilosa* flowers. The inference is that, in reverse, almost five times as much pollen from pink flowers as from white is lost to *glaberrima*. Since *P. pilosa* has a rather low number of seed set per capsule (mean number < 2), the loss is particularly significant. *P. pilosa* and *P. glaberrima* are separated by strong incompatibility barriers, although very occasionally hybrids do occur. It is almost certain, therefore, that in *P.pilosa* the white floral colour is an *ad hoc* isolating mechanism developed in response to the challenge of a sympatric species.

Species Pairs

From the consideration of sympatric and allopatric populations of a single pair of species it is only a short step to cases involving pairs of species, some sympatric and others allopatric, belonging to closely related groups. These examples are less satisfactory in that the internal control comparison is usually lacking, but where a number of comparisons can be made within a group, the results can be convincing.

An elegant demonstration of the effects of sympatry on species of such a group has been made by Grant[75] on the Leafy-stemmed Gilias. The ten species in the group are distributed along the west coast of North and South America in a manner which is ideal for a comparative study. Five of them occur in the foothills and valleys of coastal California. Each species overlaps with at least three and usually four others, with two or more species growing side by side. On the other hand four species inhabit the coastal strand. The maritime species are nowhere in contact with each other and only very rarely meet representatives of the foothill group. The tenth species, an Andean form, was not included in the study.

Twenty of the 36 possible crosses involving the nine species were completed. Wherever hybrids could be obtained they were largely sterile, so that introgression is uniformly ruled out for the whole group. The pattern of incompatibility is, however, strikingly different in the two sub-groups. The results for the various combinations are shown in table 11. Crosses between foothill species yielded a mean of 0·2 seeds per flower; while those between maritime species produced 18·1, which is about normal for intraspecific crosses. The maritime × foothill crosses gave intermediate and quite variable values.

It is most unlikely that these results can be explained by different degrees of morphological and ecological divergence. At least two of the species in the foothill group are so similar that they were at one time considered conspecific, while in the maritime group there are both diploid and tetraploid species.

It is clear, therefore, that in the Leafy-stemmed Gilias the most reasonable explanation of the difference between the sympatric group and the allopatric species results from the operation of selection in sympatry. This conclusion is strengthened by the evidence from the Cobwebby Gilias.[75] Although there are no wholly allopatric species in this group, peripheral species with allopatric populations yield signif-

TABLE 11

Comparative crossability in seeds/flower of species of Leafy-stemmed Gilias with different geographical relationships. (After Grant[75].)

Parental species	No. combinations	S/Fl in magnitude array		Mean of means, S/Fl
Foothill species *inter se*	9	0·0	0·1	0·2
		0·0	0·1	
		0·0	0·4	
		0·0	1·2	
		0·0		
Maritime species *inter se*	5	7·7		18·1
		16·7		
		19·6		
		21·9		
		24·8		
Foothill species × maritime species	8 (complete data for 6)	0·0	4·2	3·2
		0·1	5·1	
		2·8	6·8	

icantly greater numbers of seed when crossed to central species than do the central species crossed *inter se*.

In animals, there is no single example of this kind of comparison that is as complete and diagrammatic as the case of the Leafy-stemmed Gilias, but supporting evidence can be found. In one subgroup of the *Drosophila guaraní* group[98] two of the species, *D. guaraní* and *D. guarú*, occur sympatrically in Brazil while *D. subbadia* is found in Mexico, separated from the others by a large area containing no species belonging to the subgroup. In interspecific crosses only the combination of *guarú* and *subbadia* produces any hybrids at all, and in these the males are sterile. Sexual isolation is very strong between the two Brazilian species, but either of them will mate fairly readily with the Mexican species.

Similar comparisons have been made by Lorković[118] in the Alpine butterflies of the genus *Erebia*. These species are extremely similar in morphology, but cryptic divergence is indicated by the rather wide range of chromosome numbers from N = 8 to N = 51 and by the fact that hybrid inviability or sterility is found in every cross reported by Lorković. The two species in most intimate contact, *E. cassioides* and *E. nivalis* replace each other altitudinally in the eastern Alps with an overlap zone of about 100 m in vertical distance. They are well isolated ethologically. On the other hand, *E. calcarius* from the Julian Alps crosses quite easily with *E. iranica* from Persia and *E. hispania* from Spain although the chromosome numbers of these species are N = 8,

51, and 24 respectively. *E. calcarius* females will also mate with *E. cassioides* males although the reverse cross does not succeed.

Another suggestive case has been described by Dobzhansky and his colleagues in the *obscura* group of *Drosophila*.[43] They tested the mating behaviour of six species of the group by scoring the numbers of intra- and interspecific matings in an Elens-Wattiaux chamber. Three of the species, *D. pseudoobscura*, *persimilis*, and *miranda*, are closely related to one another; and the tested stocks were all collected from one locality at Mather, California. Two others, *D. bifasciata* and *imaii*, are Japanese sibling species, rather distantly related to the California species and nowhere sympatric with them. The remaining species, *D. subobscura*, is European. It meets *D. bifasciata* in Europe but does not occur in Japan where the *bifasciata* stocks were collected.

On the basis of morphology and behaviour patterns, one might expect that the three American species should cross readily among themselves, that the Japanese species should do the same, and that the members of each group should be isolated from the other and from *subobscura*. On the contrary, the closely related members of sympatric groups are just as strongly isolated from one another as they are from the more dissimilar allopatric species. In fact, the only test producing substantial numbers of interspecific matings was *D. pseudoobscura* × *imaii* (63 out of 291 matings). The results, though hardly conclusive, are consistent with the hypothesis of reinforcement of isolation in the sympatric groups.

Against these examples must be placed those cases in which sympatric species show less isolation than allopatric ones. The *Melanium* violets show such a pattern.[27] There are three partially sympatric European members of the group which cross rather easily with one another, *Viola tricolor*, *V. arvensis*, and *V. kitaibeliana*. The fourth species, *V. rafinesquii* from America, is highly incompatible with the others.

Against the Wallace Effect

The arguments which may be advanced against the reality or importance of the Wallace Effect have been most cogently stated by Moore.[132] He takes the position that since isolating mechanisms can evolve in allopatric populations there is no need to postulate a special selection for the perfection of the mechanisms. He has advanced several theoretical arguments against such selection. First, isolating mechanisms of selective origin would only be of advantage in the zone of overlap and hence would not be expected to spread through the species. If such characters were advantageous elsewhere they would simply be incorporated for their own properties and not because of their contribution to isolation. Second, if selection is to be effective in producing isolating mechanisms, then hybridization must be fairly common. As selection proceeds it becomes less and less effective since fewer hybrids are produced. The last stages of the establishment of isolation should therefore be excessively slow. Third, some sympatric species are able to form perfectly viable and fertile hybrids and yet do not do so. Fourth, examples of species which show greater isolation between allopatric than between sympatric populations may be interpreted as the results of selection for the reduction of competition, with the increased isolation resulting from the ecological differentiation of the populations. Finally, Moore suggests that in cases of severe competition, the wastage of gametes may be of advantage to the population in reducing the demands on the environmental resources.

Conclusions

The evidence for geographical variation of the degree of intraspecific reproductive isolation and the occurrence of incipient isolation in laboratory stocks indicate clearly that isolating mechanisms may develop simply as the by-products of divergence. Indeed it is logically necessary for incipient reproductive isolation of some sort to appear before selection may operate on the system. Furthermore the various experiments with artificial selection show that lowered fitness of hybrids can be effective in reducing the frequency of hybridization.

The soundest body of evidence for the Wallace Effect in nature comes from the growing number of comparisons of sympatric and allopatric populations of a species, or of pairs of species. In view of the fact that not all of the comparisons show greater isolation of the sympatric populations, it should be emphasized that evidence in the two directions is not necessarily of equal weight. It can be argued that since allopatric populations are more likely to be subject to divergent ecological conditions and are therefore likely to differ in their adaptations to the environment, the null hypothesis is not that there should be no difference between sympatric and allopatric comparisons, but that the allopatric pairs should be more reproductively isolated. That the reverse is frequently found is good evidence for the Wallace Effect. In answer to Moore's first criticism, it is clear in these cases that isolating mechanisms are not spreading from the areas of sympatry; at least the allopatric populations are lagging behind in this respect. Moreover, Crosby[210] has recently developed computer simulation models which show that some spread may be expected. As for the question of whether selection is acting to increase reproductive isolation or to reduce competition, surely these are two aspects of the same process.

A very puzzling group of observations, which illustrates another of Moore's criticisms, is the large body of data from the anurans indicating that hybrid inviability and sterility tend to follow rather than precede the development of strong behavioural isolation. There are two ways of fitting this data into the conceptual framework of the Wallace Effect. Either we have here examples of simple divergence without reinforcement of behavioural differences by selection, or else the criteria of viability and fertility are not sensitive enough to detect the adaptive inferiority of the products of hybridization. After all it must be rather subtle differences of this sort that prevent the spreading of old hybrid zones.

These hybrid zones constitute the greatest single obstacle to the acceptance of the Wallace Effect as an important phenomenon in natural populations. They demand much more intensive study than they have received in the past.

The other criticisms which Moore has levelled at the Wallace Effect are simply restatements of historical objections to the theory of natural selection which have been debated over and over again. One, concerning the reduced effectiveness of selection as reproductive isolation increases, is simply the question of the 'perfection of adaptations' first dealt with by Darwin in the *Origin of Species*.[38] Finally, the suggestion that wastage of gametes may be 'of advantage to the species' would require the elaboration of group selection to make such a scheme work at all. In view of Williams'[199] recent treatment of this theme, it hardly seems necessary to entertain the possibility.

Therefore we are left with a considerable body of evidence favouring the occurrence of selection for the improvement of isolating mechanisms, once incipient isolation has been established. The next problem to be considered is whether these initial discontinuities may not be engendered in a uniformly distributed population.

6 : Parapatric Speciation

> 'The constant elimination in each extreme region of the genes which diffuse to it from the other, must involve incidentally the elimination of those types of individuals which are most apt to diffuse. . . . The effect of such a progressive diminution in the tendency to diffusion will be progressively to steepen the gradient of gene frequency at the places where it is highest, until a line of distinction is produced, across which there is a relatively sharp contrast in the genetic composition of the species.' R. A. Fisher: *The Genetical Theory of Natural Selection.*[62]

Fisher's suggestion that reproductive isolation might arise to separate the two halves of a continuous cline has until quite recently received relatively little attention. There have been at least two developments that have served to awaken interest in the possibility. On the one hand, Mayr's[125] forceful arguments for the exclusion of any alternative to geographic speciation have paradoxically stimulated the investigation of other means of speciation. That Mayr anticipated such a response is indicated by his likening the theory of sympatric speciation to 'the Lernaean Hydra which grew two heads whenever one of its old heads was cut off'.

The second development is the growing body of observational and experimental data which is beginning to indicate the conditions under which the cohesive force of gene-flow by migration may be overcome by selection for local differentiation. A further elaboration of this theme is the possibility that the development of genetic coadaptation in areas of incipient differentiation may contribute to the differentiation itself.

Speciation by the fractionation of a continuous cline fulfils the formal definition by Mayr[125] of sympatric speciation in that it involves 'the origin of isolating mechanisms within the dispersal area of the offspring of a single deme'. It is, however, convenient to distinguish by the term parapatric speciation[175] the special case considered here *in which the incipient species are contiguous populations in a continuous cline.*

Parapatric speciation represents the most plausible of the various possibilities for sympatric speciation. Following the model proposed by Maynard Smith,[122] the establishment of the initial differentiation will be facilitated by the partial spatial separation of the populations involved. It seems inherently more likely that (1) separate density-dependent regulation of population size and (2) large selective coefficients will be found in two portions of an extended cline than in a mosaic of different environmental patches. To these conditions must be added the further strictures of either habitat selection or assortative mating or both.

We shall therefore consider the evidence for differentiation in the absence of isolation and for the development of incipient isolation, both in experimental and natural populations. We shall also consider whether the genetic structure of a clinally distributed population may contribute to this process as a force additional to that of natural selection by the external environment.

Artificial Selection and Isolation

Attempts to simulate extreme conditions of the sort anticipated by Fisher have been made with *Drosophila melanogaster*. Thoday and Gibson,[185] continuing their work with disruptive selection on sterno-pleural chaeta number, have devised a system which has in at least one instance apparently led to incipient isolation.

The flies used in this experiment were derived from a stock founded by four wild females captured in a garbage can. Four cultures were set up from this stock. In each generation, Thoday and Gibson assayed 20 males and 20 females from each of the four cultures, choosing the 8 flies of each sex with the highest and the 8 with the lowest numbers of chaetae as parents for the succeeding generation. These 32 flies were all confined in a single vial for 24 hours to allow unrestricted mating. The males were then discarded, and the females were separated into groups of four high and four low flies to yield two high and two low cultures for the next generation.

Under this regime, the lines diverged rapidly. After generation 12 the chaeta numbers of the high and low lines did not overlap, and there was no gene exchange between them. Furthermore Gibson and Thoday present evidence that the isolation came about by means of a reduction in hybridization between the high and low lines or else by competitive inferiority of the hybrids. Fig. 13a shows the distribution of chaetae in the flies assayed in generation 12. In contrast to this, Fig. 13b gives the results of forced matings carried out at the same time between high × high, high × low, low × high, and low × low flies.

The group of intermediate flies derived from the forced hybridization is conspicuously absent in the case of unrestricted mating. It is clear then that in this stock disruptive selection and frequency dependent selection have produced isolated stocks of flies with high and low

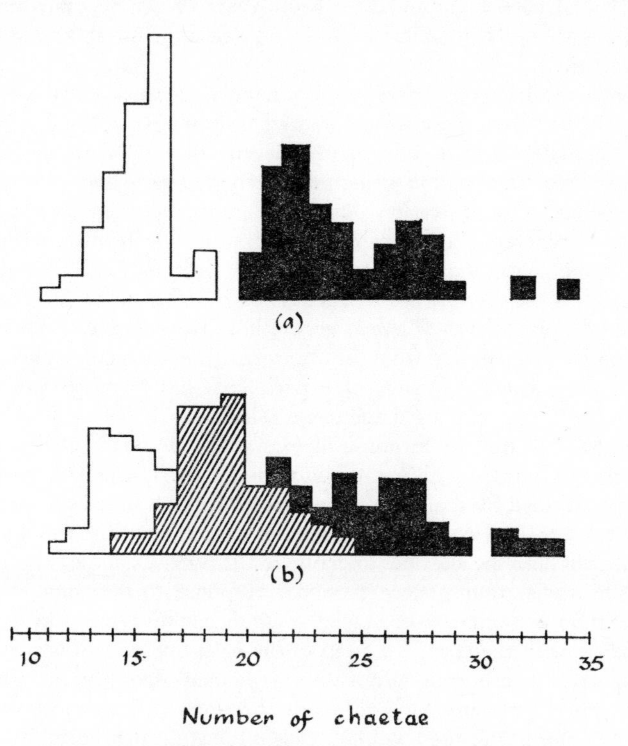

FIG. 13. The distribution of sternopleural chaeta number in lines of *Drosophila melanogaster* subjected to disruptive selection. *a*. The distribution in generation 12 when flies selected for high and low bristle number were allowed to mate without constraint. White: progeny of low females. Black: progeny of high females. *b*. The distribution derived from forced mating in generation 11. White: progeny of low × low matings. Black: progeny of high × high matings. Hatched: combined progeny of high × low and low × high matings. (After Thoday and Gibson[185].)

chaeta numbers, even though each stock is well within the 'range of dispersal' of the other.

The generality of Thoday and Gibson's result has, however, been challenged by a number of unsuccessful attempts to repeat this type of experiment.[186] Scharloo, den Boer, and Hoogmoed[164] carried out exactly the same regimen of selection, beginning with populations whose origin is better documented than that described above. Two experiments were carried out using two different stocks, each of which had been maintained for some years in laboratory culture and had shown great genetic variability in selection for several different characters.

The results were much less dramatic than those of Thoday and Gibson. Although there was an immediate response in the first generation, amounting to a difference between the high and low lines of about three bristles, the subsequent generations showed only minor divergence. The regression of the difference between high and low lines on generation number from generation one to fifteen (the last) was significant in only one of the two experiments. In neither case did complete separation of the frequency distribution of the high and low parts take place. Even as late as generation thirteen, at least one high or low parent was selected from the opposite line in each experiment.

A more extensive range of experiments has been carried out by Chabora.[16] She also used the same selection scheme as Thoday and Gibson[185] to test the response of four stocks derived from laboratory strains and four from different wild populations. The eight experiments were continued for ten generations, except for one which was lost. Two of them were carried as far as generation 42, although at generation 27 sterility of both the low lines interrupted the progress of the experiments.

Chabora's results show a limited response to selection. The 10-generation experiments produced a significant divergence in only one instance, with the change limited to the high line only. The two long-continued experiments also show significant divergences in chaeta number by generation 27, although the difference was lost with the onset of sterility in the low lines. It is interesting that in no case was a significant response elicited from a low line. In fact in seven of the eight strains, the regression of chaeta number on generation number in the low line is positive.

In both sets of experiments cited[16,164] the conclusion is drawn that no incipient isolation could be detected. Unfortunately no data are available for any of these cases on the question of incipient reproductive isolation. Information on the mating behaviour of the disruptively selected strains is badly needed.

Robertson,[212] who has carried out a similar study of disruptive selection in *Drosophila*, has discussed some of the reasons why conflicting results might be expected in experiments of this type. He points out that if mating is random and all offspring are selected from high × high and low × low crosses, a mean of only four pairs of flies contributes to each half of the population. In many generations the actual number will be less. Inbreeding, compounded by such chance effects, may reduce the vigour of the flies to the point at which crossbred flies monopolize the matings. The rather wide fluctuations of mean chaeta number in the high and low lines support Robertson's conclusion.

In addition to the experiments on chaeta number in *Drosophila*, attempts have been made to design experiments that more nearly simulate selection in natural populations for adaptation to more than one type of environment.

Robertson,[161,162] for example, has tested the ability of *Drosophila melanogaster* to adapt to a medium containing EDTA* and to retain its adaptation in the face of gene flow. The initial selection was carried out in isolated cultures and was quite successful in producing a strain capable of tolerating the EDTA medium. Table 12 shows the comparative performance of EDTA-adapted and control flies on media with or

TABLE 12

Comparative performance of EDTA-adapted and control strains of Drosophila melanogaster *in crowded or uncrowded conditions on media with and without EDTA. (After Robertson*[161]*.)*

	EDTA	Survival, ratio to controls	3 $\log_e$ thorax length	$\log_e$ larval period
Uncrowded	Nil	1·0	−7	−1
	0·0025M	1·0	−4	−15
Crowded	Nil	0·9	−14	12
	0·0025M	2·4	8	−26

without the chelating agent. The tests were performed with 0·0025M EDTA, since at higher concentrations the survival of control flies was very poor. In uncrowded conditions there is little difference in survival although the EDTA-adapted flies are consistently smaller and have a shorter developmental period on both media. Under crowded conditions, the extent of the adaptation of each strain to its environment shows up very clearly. Each strain has higher survival, larger body size, and a shorter developmental time in its appropriate medium. By manipulating the chromosomal constituents of the flies Robertson was able to show that changes in all the major chromosomes, as well as interactions between chromosomes, were involved in producing the adaptations.

* EDTA = ethylenediamine tetraacetic acid.

In order to test the maintenance of adaptation in the face of gene flow, Robertson connected pairs of cage populations with glass tubes, permitting an undetermined but presumably low rate of migration between the two types of environment. Although the results are rather variable, it appears that there was a general trend toward the reduction of the differences between the two strains. For example when the strains from the paired boxes were tested on EDTA medium and compared with the original selected strains, there was a tendency for the length of larval life to be increased in the EDTA strain and decreased in the normal strain. There has therefore been gene flow in both directions.

At generation 20, tests of assortative mating were carried out by scoring the mating behaviour of equal numbers of males and females of each strain confined in half-pint bottles. No evidence of assortative mating was found. Indeed, if there is a suggestion in the data, it is that heterogametic matings are slightly more frequent.

Robertson's final test of the differences in the two strains was to select directly for assortative mating for fourteen generations. The lines were continued using only parents which mated homogametically in choice experiments. The proportion of homogametic matings did not deviate significantly from 50% in any generation.

In contrast to these results, an experiment with houseflies by Pimentel, Smith, and Soans[150] has demonstrated the development of incipient ecological isolation in the face of substantial gene flow. The experimental system, shown in Fig. 14, was designed to exert a heavy selective pressure in favour of oviposition on either banana medium or fish medium. Ten ovipositional sites were provided in each end chamber. Only one, containing the appropriate medium, was allowed to provide offspring for the next generation. The other nine, with the alternate medium, were discarded. On the banana side, for example, eggs laid in the single banana vial were selected and those from the nine fish vials were destroyed. In the adjacent intermediate box several banana vials were placed to form a sensory barrier to flies from the fish side of the apparatus. These vials, however, were not open for oviposition. The other half of the experiment was similarly set out in reverse. Under this regime migration took place between the end chambers at a rate of about 5–15% per generation.

The populations in the two sides of the apparatus rapidly developed preferences for the banana or the fish medium. After only 18 weeks the banana flies laid 71·2% of their eggs in the banana medium, while the fish flies laid 65·2% of their eggs in the fish medium. That these preferences were the result of genetic changes in the populations was indicated by the fact that subcultures raised for two generations in the

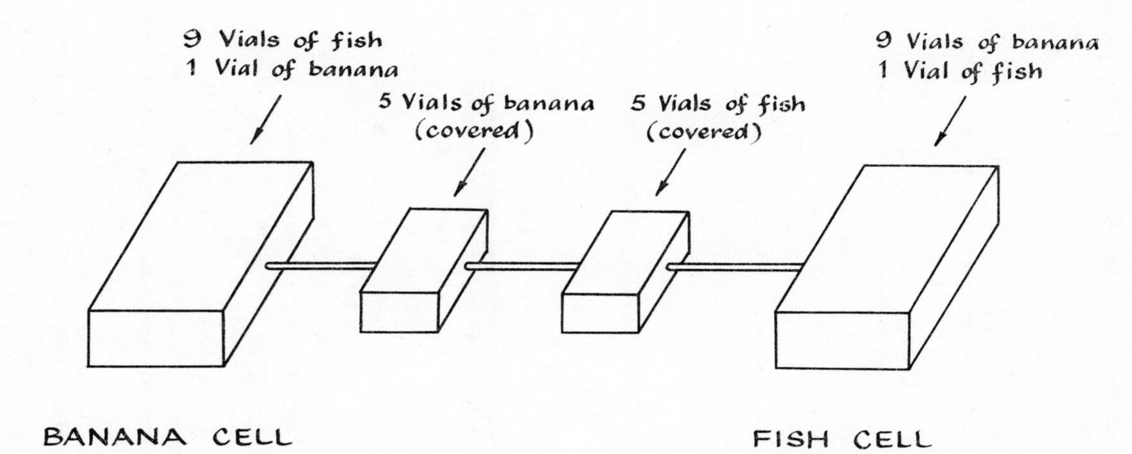

FIG. 14. Experimental chambers designed to test the development of ecological isolation in the face of gene flow. (After Pimentel, Smith, and Soans[150].)

absence of the media nevertheless still showed a strong preference for the appropriate type. Moreover in hybrids between the two strains the preference either disappeared (♀ banana × ♂ fish) or was markedly reduced (♀ fish × ♂ banana).

In this experimental system, therefore, behavioural differentiation does take place in spite of gene flow between the two extreme habitats. The important condition would seem to be the heavy selection imposed on each population. Not only does this impose lethality on all individuals represented by eggs laid on the inappropriate medium; but it also serves to reduce gene flow, since the adaptations of the flies make them susceptible to the selective 'trap' in the opposite environment.

Any attempts to assess the importance of experiments of this kind must take into consideration the likelihood that such stringent conditions may be found in nature. Until very recently it might have seemed improbable that selection gradients of this magnitude could be found, but evidence is accumulating to indicate that these simulations may not be too unrealistic.

Gradients of Selection in Natural Populations

Some of the most striking examples of the reversal of selective forces in natural populations have been described by Bradshaw and his colleagues in plants from areas of severe metal contamination. The high soil concentrations of heavy metals such as lead or copper found on mine dumps are highly toxic to most plants. Some species such as the grass *Agrostis tenuis*, however, can develop tolerant strains which are capable of invading these inhospitable habitats. Furthermore, the differentiation of tolerant strains may take place in relatively tiny areas immediately adjacent to populations of non-tolerant individuals without any restriction of pollen exchange between them. The tolerance is heritable and not induced by a conditioning process.

Bradshaw[7,91] investigated, for example, the population of *Agrostis tenuis* on the Goginan lead mine near Aberystwyth in Wales. The mine covers a very limited area, and the lead content of the soil drops off very rapidly to normal pasture. Bradshaw collected a series of samples along a transect across the edge of the mine and out into the pasture. From each station twenty tillers were gathered, grown in a garden in randomized blocks for two years, and then tested for lead tolerance. Tolerance was scored by expressing the amount of growth in a lead-containing solution as a percentage of growth in a normal solution. Fig. 15 shows the soil concentration of lead and the index of lead tolerance along the transect. Over about ten metres across the boundary there is a striking change in the degree of tolerance of the plants.

Perhaps the most surprising aspect of the results is that the two samples on either side of the boundary are as extreme as those considerably further away.

Jain and Bradshaw[91] have calculated coefficients of selection against the unadapted types at the Goginan site by comparing the growth of spaced plants on the two kinds of soil. They estimate the coefficient of selection against the tolerant plants in the pasture habitat as 0·40 and that against the non-tolerant plants on the mine as 0·95. These selective forces should be adequate to maintain the sharp boundary between the two parts of the population even if pollination and seed dispersal are unrestricted.

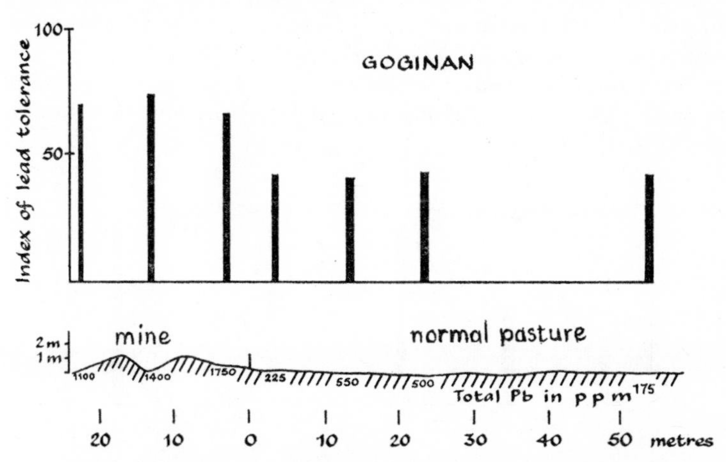

FIG. 15. The pattern of differentiation in the populations of *Agrostis tenuis* in response to marked changes in the concentration of lead in the soil. (After Jain and Bradshaw[91].)

There are indications, however, that incipient isolating mechanisms are forming between the tolerant and non-tolerant plants. McNeilly and Antonovics[126] have studied the compatibility relationships and flowering times of *Agrostis tenuis* and *Anthoxanthum odoratum* along transects from mine dumps to uncontaminated pastures. In both species the tolerant populations from the mines flowered significantly earlier than the adjacent populations from the pastures. In *Anthoxanthum* there was also a suggestion of incompatibility in that non-tolerant plants set fewer seeds per plant when pollinated by tolerant plants.

Another and perhaps more general example of differentiation in the face of gene flow may be cited from the work of Aston and Bradshaw[2] on the growth of *Agrostis stolonifera* in maritime habitats. In this species the length of the stolons varies according to the degree of exposure to winds and storms. The differences are genetic since they are maintained in garden culture, where the distributions of stolon length in cliff-face stocks and in sheltered-pasture stocks do not even overlap. Aston and Bradshaw have shown that the differences are adaptive. The cliff plants have a high mortality when grown in wet, heavy soil, the coefficient of selection being of the order of 0·50. They are quite unable to survive under constant flooding. The long stolon plants, however, can survive flooding, even if they come from relatively dry areas. Thus stolon length is the character of importance here. In exposed conditions, the plants with short stolons survive undamaged, while those with long stolons are much battered by wind. Jain and Bradshaw[91] estimate the coefficient of selection against the plants with long stolons in such conditions to be about 0·80.

Populations of *Agrostis stolonifera* may show extremely abrupt transitions where sudden changes of habitat occur. Fig. 16 shows the

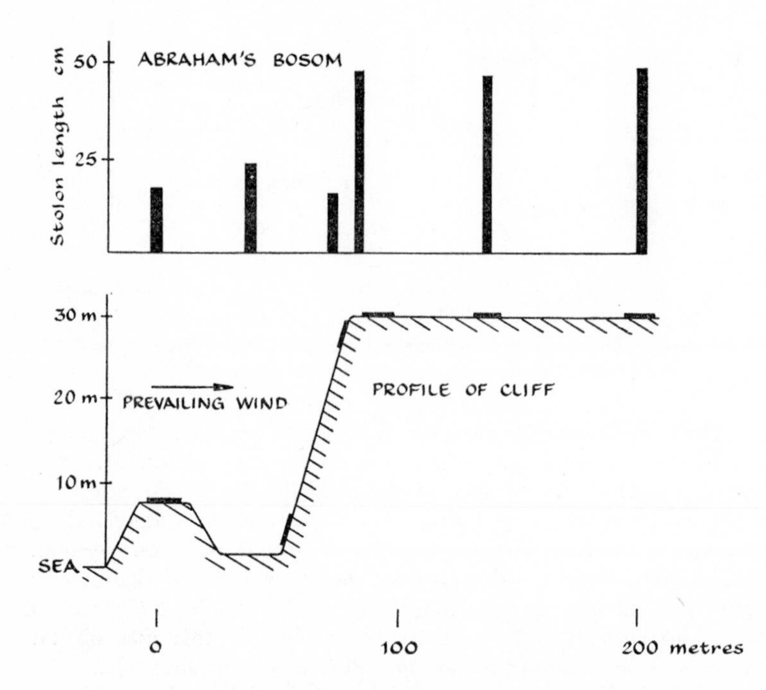

FIG. 16. The pattern of differentiation in populations of *Agrostis stolonifera* in response to variation in exposure to storm winds. (After Jain and Bradshaw[91].)

results of a transect collected by Aston and Bradshaw on sea cliffs at Abraham's Bosom in Wales. Note that the stolon lengths represented in this diagram are not those of the populations *in situ* but are the results of growth under garden conditions. The transition is complete in less than ten metres with the adjacent samples as extreme as any in the transect.

Because of the prevailing winds blowing from the sea, gene flow in this case is almost unidirectional. Seedlings from both the cliff populations and the inland populations are smaller than the parental generation, indicating that the movement of pollen is largely inland. The inland seedlings are also much more variable than the surviving adult population.

These effects in species of *Agrostis* are remarkable principally because of the extremely small scale of the differentiation in space. The same pattern may be found frequently over much larger areas, and in favourable circumstances the selective agent has been identified. A case in point is that of the butterflies of the *Limenitis arthemis-astyanax* complex. The relationship of these forms is of particular interest as they were regarded by Fisher[66] as incipient species in the process of developing reproductive isolation without geographical isolation. In colour, the forms known as *arthemis* (the White Admiral) and *astyanax* (the Red-spotted Purple) are strikingly different from one another. *Arthemis* has a broad white band across the fore and hind wings, and there is a row of reddish spots between the band and the marginal iridescent spots of the hind wings. In *astyanax* the white band is missing and the reddish spots are greatly reduced or absent. A further difference between the two is the suffusion of blue-green iridescence found on the hind wings in *astyanax* but not in *arthemis*. The white band appears to be controlled by a single incompletely recessive gene while the other two characters show continuous variation.[151,158]

Platt and Brower[151] have studied the geographical distribution of the three characters in seven populations forming a transect from Quebec to Virginia. In the extreme northern samples the white band is virtually fixed; in the south it is absent. In the area of southern New England and the northern mid-Atlantic states there is a transition zone characterized by intermediate forms and high variance. The pattern of change in the other characters is similar, although the variance of these in all northern populations is high.

The pattern of *astyanax*, differing as it does from *arthemis* and from other closely-related species of *Limenitis*, has long been interpreted[154] as a response to selection for mimicry of the swallowtail *Battus philenor*. This butterfly, which has been shown to be extremely distasteful to birds,[8] is faithfully copied on both dorsal and ventral sides by *astyanax*.

Furthermore, Platt and Brower have shown that the zone of transition between *arthemis* and *astyanax* closely parallels the northern limit of the distribution of *Battus philenor*. The selective value of the white-banded pattern is less clear, but it has been suggested that it serves to provide a disruptive and therefore a confusing image to potential predators.

The question of the evolutionary status of these forms has been much debated. Remington[159] has concluded that *arthemis* and *astyanax* represent formerly disjunct populations which have recently rejoined and are now hybridizing with one another. Platt and Brower,[151] however, have argued that there is no evidence of former allopatry and no indication of incipient speciation. They cite evidence that breeding is random with respect to colour, that crosses show no hybrid inviability, sterility, or disturbance of the sex-ratio, and that the genitalia of the two forms do not differ in any consistent fashion. Thus *Limenitis arthemis* provides an example of a cline between the two areas of opposed selective forces, differing from the previously mentioned examples only in the geographical scale of the cline.

The Reinforcement of Clinal Changes

In the cases cited above, the zone of maximum phenotypic change or the 'step' in the cline coincides with a major environmental discontinuity which in itself produces a reversal of selective value. There is, however, reason to believe that such steps may arise along the course of a smooth environmental gradient. Clarke[21] has investigated the properties of a model leading to the steepening of clines without a corresponding change of ecological gradients.

The model proposes a series of populations distributed along a smooth gradient of some environmental factor which affects the selective value of the genotypes at one locus. If the polymorphism at this locus is maintained by heterozygous advantage and if the selective values are proportional to the position of the population along the gradient then the genotypes and their selective values may be written as:

Genotypes	A_1A_1	A_1A_2	A_2A_2
Selective value	$1-d$	1	d

where d is a function of distance. According to Fisher's[62] expression for the equilibrium value of A_2, where a, b, and c are the selective values of A_1A_1, A_1A_2, and A_2A_2:

$$\hat{q} = \frac{b-a}{2b-a-c} = d$$

indicating a smooth cline in the frequency of A_2 along the environmental gradient. This smooth change may be altered by the introduction of a modifier at a second locus which differentially affects the genotypes at the first. For the sake of simplicity, the modifier is assumed to be dominant, the recessive homozygote having no effect on fitness. If r, s, and t represent the degrees of interaction with the genotypes of the first locus, then the fitness of the various combinations becomes:

	A_1A_1	A_1A_2	A_2A_2
B_1B_1	$1-d$	1	d
B_2-	$1-d+r$	$1+s$	$d+t$

B_2 will spread whenever the average fitness of B_2- genotypes is greater than that of B_1B_1 genotypes, *i.e.* when:

$$p^2(1-d+r)+2pq(1+s)+q^2(d+t) > p^2(1-d)+2pq+q^2d$$

or

$$q^2(r+t-2s)+2q(s-r)+r > 0.$$

If B_2 becomes fixed then, again from Fisher's expression:

$$\hat{q} = \frac{d+s-r}{1+2s-r-t}$$

so that the slope of the cline is:

$$\frac{1}{1+2s-r-t}.$$

The effect of the B_2 gene on the cline can then be seen to depend on the relative size of s, r, and t. The cline will become steeper if $2s < r+t$, an effect equivalent to disruptive selection. If $2s > r+t$, however, the cline will be less steep as a result of stabilizing selection.

Unless all three coefficients are positive B_2 will not become fixed, but will be limited in its spread by the value of q. The result will be the insertion of steps into the cline at those points where B_2-bearing genotypes lose their advantage. The most likely case is that in which selection is reversed at the two ends of the cline, *i.e.* with either r alone, or r and s, negative. Solving the quadratic above, B_2 spreads when:

$$q > \frac{s-r-\sqrt{s^2-rt}}{2s-r-t}$$

and a step develops in the cline where:

$$d = \frac{s-r-\sqrt{s^2-rt}}{2s-r-t}.$$

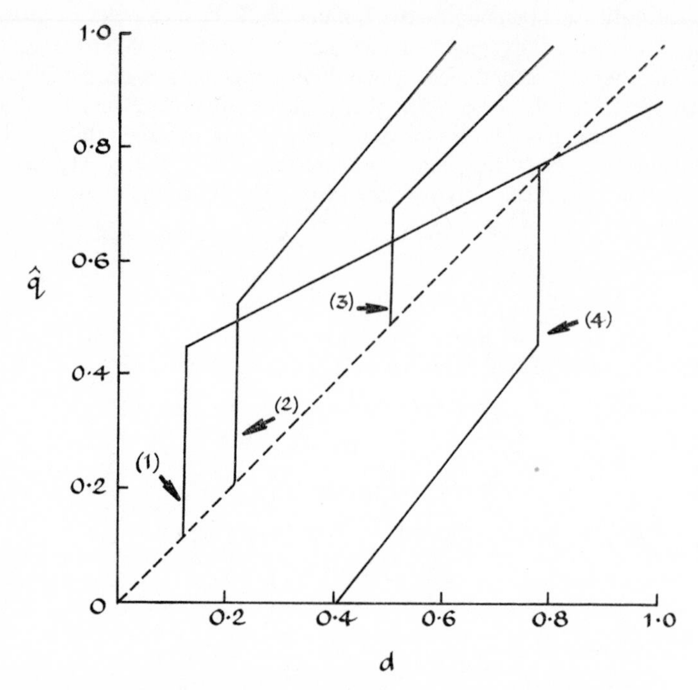

FIG. 17. The effect of a modifier on a smooth cline of gene frequency. The interaction coefficients (see text) are: (1) $r=-0{\cdot}20$, $s=0{\cdot}60$, $t=0{\cdot}40$; (2) $r=-0{\cdot}10$, $s=0{\cdot}10$, $t=0{\cdot}50$; (3) $r=-0{\cdot}20$, $s=0$, $t=0{\cdot}20$; (4) $r=0{\cdot}50$, $s=0{\cdot}10$, $t=-0{\cdot}10$. The solid lines indicate deviations induced by the modifier in each case. $\hat{q}$=equilibrium gene frequency; d=relative distance along the cline. (After Clarke[21].)

The effects on the cline for various values of r, s, and t are shown in Fig. 17.

The important aspect of the model is that the area of rapid change of gene frequency does not correspond to any discontinuity in the environment. Although there is of course an ultimate dependence of the change on an environmental gradient, the step is the result of a selective interaction between components of the genetic system.

Area Effects

In many polymorphic species, the occurrence of rather sharp transitions between regions characterized by quite different gene frequencies is a common phenomenon. The term *area effect* has been coined by Cain and Currey[11] to describe some striking examples of this kind of distribution in *Cepaea nemoralis.*

Populations of these snails in central England characteristically show one of two patterns of variation. On the one hand there is the classic condition studied by Cain and Sheppard[13,14] in which the composition of populations is rather closely correlated with the visual properties of the environment in which the snails are living. Dark woodland habitats tend to harbour high proportions of pink and brown unbanded types, while greener and less uniform habitats have higher proportions of yellow banded snails. The transitions from place to place may be quite abrupt, but they coincide with the discontinuities of the habitats. This pattern is found in those parts of the country in which the principal avian predator of *Cepaea,* the Song Thrush (*Turdus ericetorum*), is common (see also [35]).

In contrast to the rather small-scale pattern formed by this adjustment of populations to the local environment, there are regions such as the Marlborough Downs where the distribution of phenotypes is very different. Cain and Currey[11] have described the occurrence of an area of about four by seven kilometres over which the five-banded phenotype is virtually absent. From 103 collecting sites only 11 out of 5767 snails of this normally common type were collected. The populations consisted almost entirely of unbanded or midbanded individuals. The latter phenotype is determined by an unlinked modifier of the principal locus determining banded versus unbanded shells. Since it is dominant to the five-banded condition but unexpressed in unbanded individuals, the occurrence of large numbers of midbanded individuals shows that the population is virtually saturated with the modifier. Hence in this area there is an unusually high frequency of both the gene for absence of bands and that for modification of the banding pattern.

The most surprising aspect of this unusual situation is that to the northeast of this area there is a complete transition over less than two kilometres to populations consisting of up to 100% five-banded individuals, with both unbanded and midbanded totally absent. These relationships are illustrated in Fig. 18.

In the same region, similar effects are found in the distribution of shell colours. In the southwestern portion of the Downs there is an area of about nine square kilometres characterized by high frequencies of the brown phenotype. In some places this normally scarce type may make up 98% of the population. To the north the frequency of browns changes from 73% to 12% over a distance of about 130 metres, with no evidence of any isolation between populations.

Cain and Sheppard[14] have devised a method of displaying on a scatter diagram the relation of *Cepaea nemoralis* populations to their

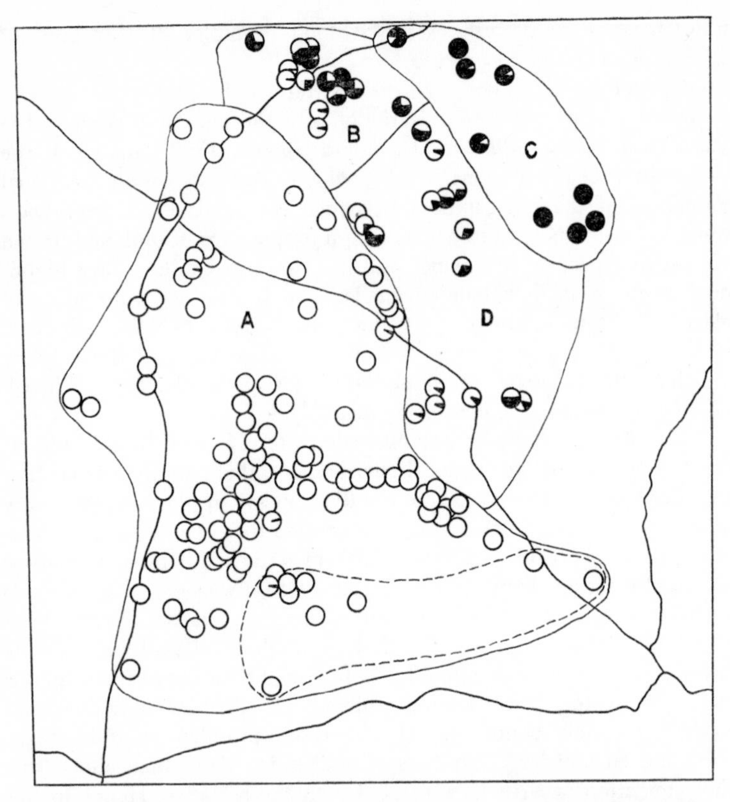

FIG. 18. Area effects in *Cepaea nemoralis* on the Marlborough Downs. The proportion of five-banded shells in each sample is indicated by the blackened sector. In area A the populations consist almost entirely of unbanded and midbanded individuals. In area *C* these two dominant genes are virtually absent. (After Cain and Currey[11].)

habitats. By plotting the proportion of yellow shells in a population against the proportion of effectively unbanded shells (*i.e.* with at least the top two bands missing) they have shown that in populations from regions where visual predation is an important factor there is distinct grouping of habitats of similar composition. Fig. 19 shows such a plot for colonies from the countryside within ten miles of Oxford. In striking contrast to this pattern, Fig. 20 shows samples from north-eastern and southwestern portions of the Marlborough Downs. It is evident that here there is no suggestion of the previous pattern. The two areas are distinct, but the area effects completely override any influence of different habitats within the areas.

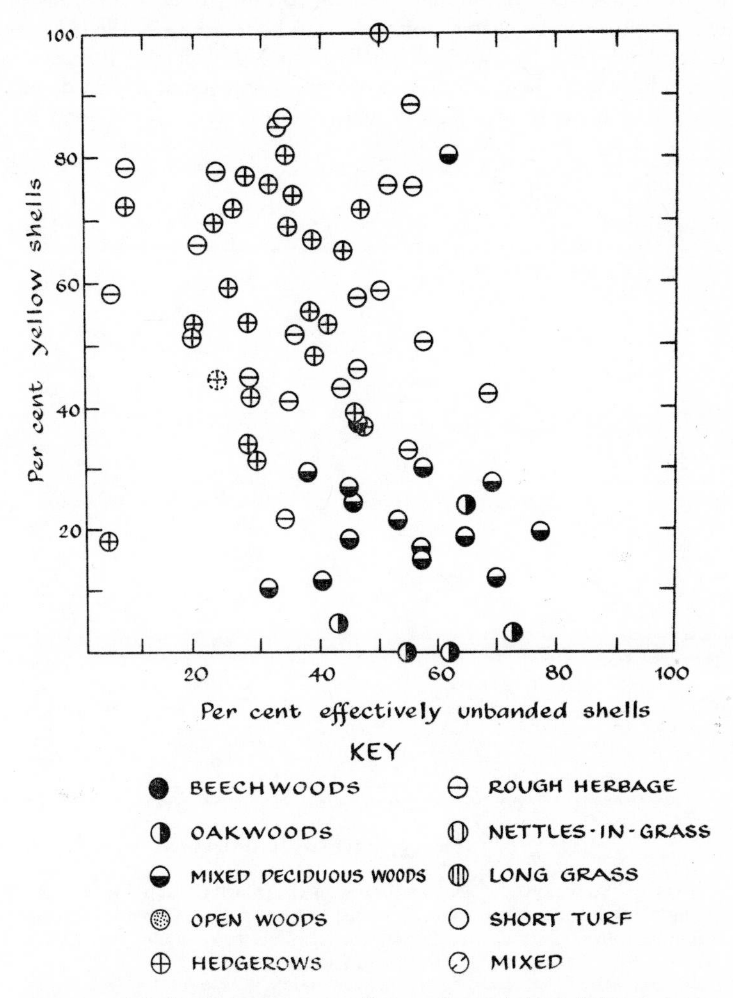

FIG. 19. Variation in *Cepaea nemoralis* according to habitat. The relationship between the proportion of yellow shells and the proportion of effectively unbanded shells (*i.e.* with bands 1 and 2 absent) in colonies within 10 miles of Oxford. (After Cain and Currey[11].)

Cain and Currey[11] have presented evidence for great stability of the area effects of the Marlborough Downs. Not only does the past history of agriculture argue against any widespread alterations in the character of the landscape, but subfossil shells from Neolithic and Bronze Age sites adjacent to the modern area of high frequency of brown demonstrate the existence of a high brown area as early as about 1000 B.C.

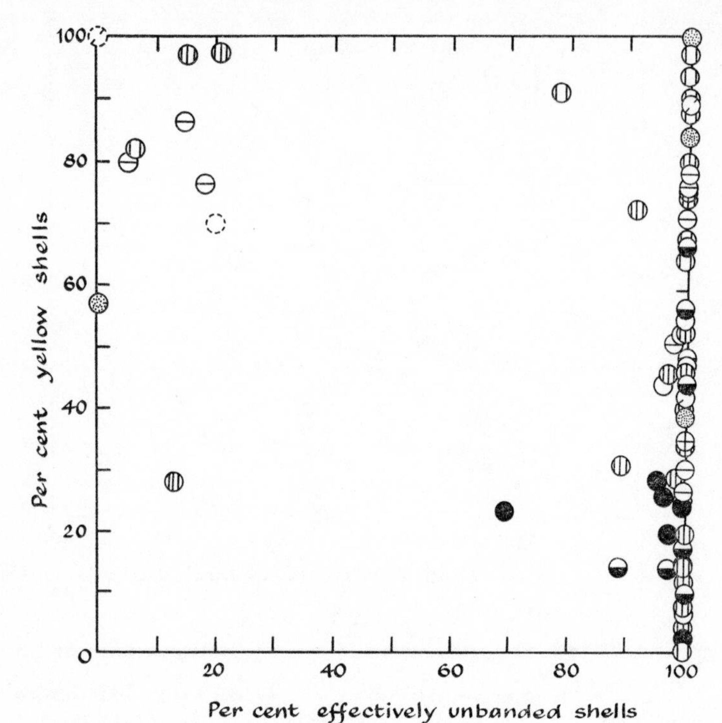

FIG. 20. Variation in *Cepea nemoralis* on the Marlborough Downs. On the right-hand side are plotted samples from area A. On the left are samples from area *C*. See figure 6.6. Symbols as in figure 6.7. (After Cain and Currey[11].)

Thus it seems unlikely that these area effects are the results of 'genetic bottlenecks' or 'founder effects', as Goodhart[73] has suggested.

Cain and Currey[11,12] have argued that the area effects on the Marlborough Downs are caused by subtle differences from place to place in the direction and strength of environmental selection. They suggest, for example, that the brown phenotype may be better able to withstand low temperatures than other morphs. The geographic distribution of brown in *C. nemoralis* tends to support this view since the type is rare in France but increases in frequency in northern Europe and Great Britain. There also seems to be a correlation on the Marlborough Downs between the occurrence of high frequencies of brown and the presence with *nemoralis* of *C. hortensis*, a species which ranges much further north than *nemoralis*. High frequencies of brown are found in valley situations which would collect the cold air flowing downwards from the Downs.

This interpretation has been questioned by Clarke[22] on the basis of evidence collected in Scotland where *C. nemoralis* reaches its northern limit. If the brown morphs were significantly better adapted to cold climates then high frequencies should be found in these northern populations. It turns out, however, that although area effects are found, the brown morph does not contribute to them. Since in the other cases of area effects on the Marlborough Downs there is not even a suggestion of a selective agent, the case for direct environmental effects is far from established.

Clarke[22] has argued for the interpretation of area effects as examples of clinal 'steps' resulting from coadaptations of the sort postulated in the model discussed above. In the case of *Cepaea* the inference is rather indirect, but in another group of polymorphic land snails there are strong indications of the effects of the genetic environment. Following the pioneering work of Crampton[31] on the snails of the genus *Partula* from the Polynesian island of Moorea, Clarke and I undertook a study of the distribution of species and morphs with respect to the physical and biotic factors of the environment.[22,23,138,139]

Among the Moorean species of *Partula*, the least complex association is found in the northwestern portion of the island where only two well-differentiated species occur, *P. taeniata* and *P. suturalis*. *P. taeniata* is particularly common in this region. It is found at altitudes greater than 500 metres on the ridges and descends in at least one place to within five metres of the sea. In favourable habitats the density may rise to more than 10 individuals per square metre, although 1 to 2 per square metre is more usual.

The scale of the variation in two of the characters for which *P. taeniata* is polymorphic is indicated in Figs. 21 and 22. Each circle represents a random sample from a limited area, usually 10 × 10 metres. The two forms represented, banding of the shell and 'purple' shell colour, are both dominant.[138]

Fig. 21 shows that bandedness is nowhere very common in the area. Usually there are less than 10% banded shells, although in the two extreme western samples the frequency rises to 20%. The distribution

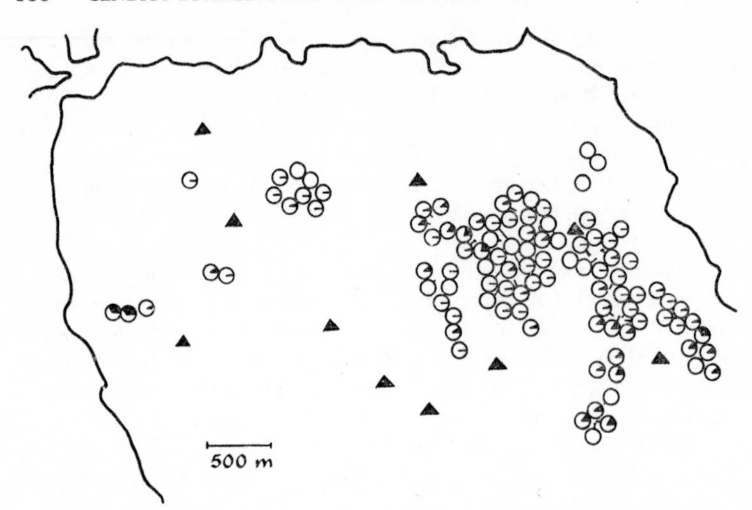

FIG. 21. *Partula taeniata* in northwestern Moorea. The proportion of banded shells in each sample is indicated by the blackened sector, where 6 degrees of arc represent 1 per cent. Black triangles show mountain peaks. (After Clarke[22].)

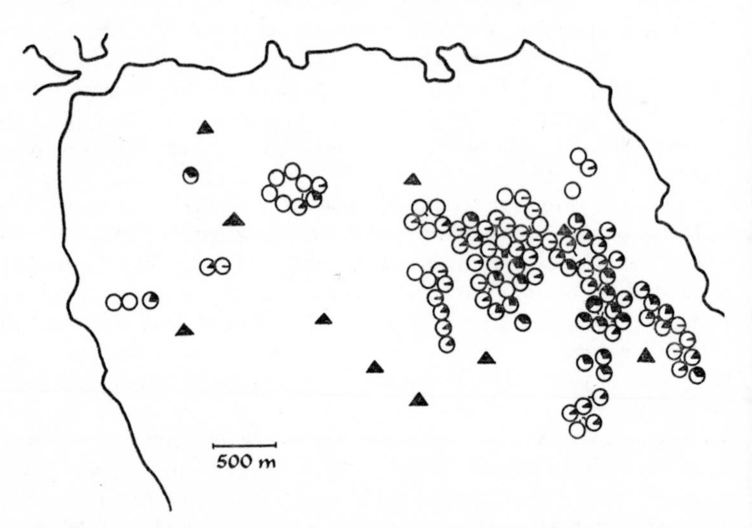

FIG. 22. *Partula taeniata* in northwestern Moorea. The proportion of 'purple' shells (N3+N4 of Murray and Clarke, 1966) in each sample is indicated by the blackened sector, where 4 degrees of arc represent 1 per cent. The small square shows the location of the centre of the transect. (After Clarke[22].)

of purple shells, shown in Fig. 22, encompasses a greater range of frequencies, exceeding 30% in the eastern portion of the region. Here the characteristic pattern of an area effect may be observed, since the frequencies drop off very rapidly at the periphery.

Attempts to interpret the patterns of variation in terms of environmental variables have proved unsuccessful. The area of high frequency of purple, for example, extends across the ridge separating two major valleys, embracing slopes which differ in aspect and hence in temperature, rainfall, and vegetation. The analysis of the distribution of plants, which might serve as ecological indicators, shows significant relations of groups of plants but no correlations with the phenotypes of the snails. Moreover the boundary of the high purple area occurs in an area where the ecological conditions on the two sides are much more similar than are the extreme habitats within the area.

On the other hand, there are strong indications that natural selection is in fact operating on these loci. In laboratory crosses between heterozygotes for the gene for banding, there is a marked deficiency of dominants among the offspring (see table 13), indicating that the

TABLE 13

Offspring of matings between heterozygotes for the gene for banding in Partula taeniata, *showing a significant deficiency of banded individuals.*

	Offspring from 21 Matings		Deviation from 3:1 Ratio	Heterogeneity among Matings
Banded	606	χ^2	4·86	16·51
Unbanded	239	d. f.	1	20
Total	845	P	<0·05	0·70–0·50

homozygous dominants are subvital and that therefore there must be counterselection in nature to support the observed frequencies of bandeds.

In order to investigate the transition from one area effect to another, Clarke and I have sampled the border by means of a transect of contiguous 10 metre squares stretching 200 metres from a sample containing 37% purple shells to one with 0%. Fig. 23 shows the proportions of both purple shells and banded shells along the transect. The number of animals in each sample is given in the box below.

The transect shows several rather surprising characteristics. First, in an area which does not appear to be unsuitable habitat, the density of individuals declines to approximately zero. Second, although there are very significant differences in the frequency of both types along the transect, the major changes do not coincide with the reduction of population size. The major decrease in purple takes place in the thickly

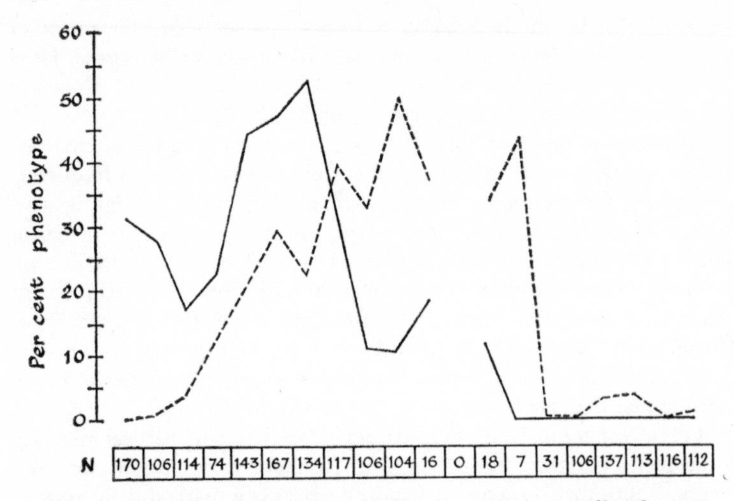

FIG. 23. A transect of populations of *Partula taeniata* extending from an area of high frequency of 'purple' shells to one of low frequency. The solid line shows the proportion of purple shells in each 10-metre square, while the dashed line shows the proportion of banded shells. The boxes give the numbers of adult snails in each square. (After Clarke[22].)

populated area on the 'purple' side of the break, while the drastic drop in bandedness is on the other side. Third, there is a quite unsuspected and extraordinary increase in the frequency of bandeds in the centre of the transect. In no other sample from this whole region does the frequency even approach that seen here, approximately 50%. This increase is accompanied by a loss of distinctness of the banded phenotype. Some of the shells are extremely difficult to score, suggesting a breakdown of dominance of the allele for banding.

At least two other observations point up the disharmonious nature of this zone of transition. There is an inverse correlation between the proportion of purple shells in any sample and the size of the shells in that sample. Within samples, however, there is no relation between colour and size. In other words, although the mean size of purple shells is the same as that of shells of other colours, nevertheless the more purple shells there are in any sample, the smaller the mean size of the sample.

A final peculiarity of these samples is the variation in fecundity along the transect. *Partula* species are ovoviviparous hermaphrodites, producing offspring one at a time at fairly regular intervals. Hence an examination of the number of uterine eggs and young provides an index of breeding condition and fecundity. Along the transect, the mean number of eggs and young rises significantly from the high purple to the low purple area. Part of this effect may be ascribed to a positive correlation between size and number of embryos, but the increase in size in the low purple area is not great enough to explain the whole of the difference.

We therefore have evidence of a gross disturbance of the genetic system of *Partula taeniata* on the border of an area effect, including (1) large changes in the frequency of major alleles at two loci, (2) changes of population density, (3) changes in body size which are related to the colour composition of the population but not to the colour of individuals, (4) disturbances of dominance relationships, and (5) variations in fecundity. Changes of this sort are not to be expected as the result of a direct effect of a clinal change in an environmental selective agent, nor even as the result of a complete reversal in the direction of selection. Clarke[22] has argued that the difficulties are largely dissipated if the border is considered to be a transition zone between two areas with different systems of coadaptation. The development and reinforcement of a step in a previously clinal distribution results in the concentration of the phenomena of genetic readjustment. It is perhaps reasonable to apply the term 'primary hybridization' to this process, implying that many of the phenomena are similar to those which occur when previously isolated populations regain contact with one another.

Conclusions

This considerable body of evidence, although it is derived by a variety of methods applied to many different organisms, does not yet permit us to draw an unequivocal conclusion on the likelihood of parapatric speciation. The broad outlines of the problem, however, are clear enough to show the gaps in our knowledge. It has been established, both by laboratory experiments with flies and by the observation of clines in nature, that significant differentiation can take place in the face of massive gene flow, provided that the conditions of strong selection and independent regulation of population size are met. It is also highly likely that the architecture of the genetic system may reinforce and enhance the differentiation of coadapted gene complexes.

Since we have already seen in an earlier chapter that selection may lead to the perfection of isolating mechanisms between incipient species, it becomes apparent that the stage in the process about which very little is known is the intermediate one between differentiation and selection for isolation. The conflicting results obtained in different laboratories from experiments on disruptive selection, the lack of success in selecting for isolation *per se*, the existence of and the stability of stepped clines all point to a great inertial block at this intermediate level. If we are to gauge the importance of parapatric speciation as a means of increasing the number and diversity of species without the necessity for geographic isolation, it may be necessary to extend the length of time usually devoted to selection experiments in the laboratory or to devise new methods of attacking the problem in natural populations.

Prospect

A century after the publication of *On the Origin of Species*, Darwin's assessment of the extent of genetic variation in natural populations has been finally vindicated. First by means of indirect experiments and then by more direct demonstration of biochemical polymorphisms, it has become abundantly clear that the diversity within the gene pool of many species is very great.

It is also evident that a large part of the genetic variation is influenced by natural selection. The patterns of variation within and between species do not suggest random fluctuations in the frequency of neutral alleles, and the maintenance of variation in the face of reduced population size argues that alleles are not greatly affected by genetic drift.

While it is possible that much of the observed polymorphism could be maintained by selection in favour of heterozygotes, a major alternative exists in frequency-dependent selection. If the fitness of phenotypes depends on their frequency in a population, then a whole new range of stable equilibria is possible. Non-random mating, predator-prey interactions, disease, parasitism, and intraspecific competition provide opportunities for the operation of frequency-dependent selection.

An unusual property of polymorphisms maintained by frequency-dependent selection is their stability even though the heterozygote may be at a disadvantage to both homozygotes. The selective loss entailed by such selection against the heterozygotes may be reduced in two ways. Either the phenotypic expression of the heterozygote may be altered by the evolution of dominance or the heterozygotes may be eliminated by the development of reproductive isolation.

The tremendous genetic diversity of natural populations has only just begun to be explored. We need to know more about the mathematical implications of this rich store of variation, especially with regard to the consequences of frequency-dependent selection. We also need to know much more about selection and speciation in natural populations. Only by these means can the apparent contradictions of current evolutionary theory be resolved.

References

1. Allen, J. A. and Clarke, B. 1968. Evidence for apostatic selection by wild passerines. *Nature*, **220**, 501–502.
2. Aston, J. L. and Bradshaw, A. D. 1966. Evolution in closely adjacent plant populations. II. *Agrostis stolonifera* in maritime habitats. *Heredity*, **21**, 649–664.
3. Bauer, K. 1957. Zur systematischen Stellung des Blutspechtes. Johann Friedrich Naumann-Ehrung: 22–24. Deutscher Kulturbund, Berlin.
4. Blair, W. F. 1964. Isolating mechanisms and interspecies interactions in anuran amphibians. *Q. Rev. Biol.*, **39**, 334–344.
5. Bodmer, W. F. and Edwards, A. W. F. 1960. Natural selection and the sex ratio. *Ann. hum. Genet.*, **24**, 239–244.
6. — and Parsons, P. A. 1960. The initial progress of new genes with various genetic systems. *Heredity*, **15**, 283–299.
7. Bradshaw, A. D. 1960. Population differentiation in *Agrostis tenuis* Sibth. III. Populations in varied environments. *New Phytol.*, **59**, 92–103.
8. Brower, J. V. Z. 1958. Experimental studies of mimicry in some North American butterflies. Part II. *Battus philenor* and *Papilio troilus*, *P. polyxenes* and *P. glaucus*. *Evolution*, **12**, 123–136.
9. — 1960. Experimental studies of mimicry. IV. The reactions of starlings to different proportions of models and mimics. *Am. Nat.*, **94**, 271–282.
10. Brower, L. P. and Brower, J. V. Z. 1962. The relative abundance of model and mimic butterflies in natural populations of the *Battus philenor* mimicry complex. *Ecology*, **43**, 154–158.
11. Cain, A. J. and Currey, J. D. 1963a. Area effects in *Cepaea*. *Phil. Trans. R. Soc. B*, **246**, 1–181.
12. —— 1963b. The causes of area effects. *Heredity*, **18**, 467–471.
13. — and Sheppard, P. M. 1950. Selection in the polymorphic land snail *Cepaea nemoralis*. *Heredity*, **4**, 275–294.
14. —— 1954. Natural selection in *Cepaea*. *Genetics*, **39**, 89–116.
15. —— 1957. Some breeding experiments with *Cepaea nemoralis* (L.). *J. Genet.*, **55**, 195–199.
16. Chabora, A. J. 1968. Disruptive selection for sternopleural chaeta number in various strains of *Drosophila melanogaster*. *Am. Nat.*, **102**, 525–532.
17. Clarke, B. 1960. Divergent effects of natural selection on two closely-related polymorphic snails. *Heredity*, **14**, 423–443.

18. CLARKE, B. 1962a. Balanced polymorphism and the diversity of sympatric species. *Publs Syst. Ass.*, **No. 4,** 47–70.
19. — 1962b. Natural selection in mixed populations of two polymorphic snails. *Heredity*, **17,** 319–345.
20. — 1964. Frequency-dependent selection for the dominance of rare polymorphic genes. *Evolution*, **18,** 364–369.
21. — 1966. The evolution of morph-ratio clines. *Am. Nat.*, **100,** 389–402.
22. — 1968. Balanced polymorphism and regional differentiation in land snails. In: *Evolution and Environment*, ed. Drake, E. T. Yale University Press.
23. — and MURRAY, J. 1969. Ecological genetics and speciation in land snails of the genus *Partula*. *Biol. J. Linn. Soc.*, **1,** 31–42.
24. — and O'DONALD, P. 1964. Frequency-dependent selection. *Heredity*, **19,** 201–206.
25. CLARKE, C. A. and SHEPPARD, P. M. 1960. The evolution of dominance under disruptive selection. *Heredity*, **14,** 73–87.
26. — — 1962. The genetics of the mimetic butterfly *Papilio glaucus*. *Ecology*, **43,** 159–161.
27. CLAUSEN, J., CHANNELL, R. B., and NUR, U. 1964. *Viola rafinesquii*, the only *Melanium* violet native to North America. *Rhodora*, **66,** 32–46.
28. COOK, L. M. 1965. A note on apostasy. *Heredity*, **20,** 631–636.
29. — and MURRAY, J. 1966. New information on the inheritance of polymorphic characters in *Cepaea hortensis*. *J. Hered.*, **57,** 245–247.
30. CORRENS, C. 1928. Bestimmung, Vererbung und Verteilung des Geschlechtes bei den höheren Pflanzen. *Handb. VererbWiss.*, **Band II,** 1–138.
31. CRAMPTON, H. E. 1932. Studies on the variation, distribution and evolution of the genus *Partula*. The species inhabiting Moorea. *Publs Carnegie Instn*, **No. 410,** 1–335.
32. CROSBY, J. L. 1963. The evolution and nature of dominance. *J. Theoret. Biol.*, **5,** 35–51.
33. — 1966. Self-incompatibility alleles in the population of *Oenothera organensis*. *Evolution*, **20,** 567–579.
34. CROW, J. F. 1961. Population genetics. *Am. J. hum. Genet.*, **13,** 137–150.
35. CURREY, J. D., ARNOLD, R. W., and CARTER, M. A. 1964. Further examples of variation of populations of *Cepaea nemoralis* with habitat. *Evolution*, **18,** 111–117.
36. DAMIAN, R. T. 1964. Molecular mimicry: Antigen sharing by parasite and host and its consequences. *Am. Nat.*, **98,** 129–149.
37. DARLINGTON, C. D. 1958. *Evolution of Genetic Systems*, 2nd ed. Oliver and Boyd, Edinburgh.
38. DARWIN, C. 1859. *On the Origin of Species*. Murray, London.
39. — 1874. *The Descent of Man*, 2nd ed. Murray, London.

40. Diver, C. 1929. Fossil records of Mendelian mutants. *Nature*, **124**, 183.

41. Dobzhansky, Th. 1941. *Genetics and the Origin of Species*, 2nd ed. Columbia University Press, New York.

42. — 1948. Genetics of natural populations. XVIII. Experiments on chromosomes of *Drosophila pseudoobscura* from different geographic regions. *Genetics*, **33**, 588–602.

43. — Ehrman, L., and Kastritsis, P. A. 1968. Ethological isolation between sympatric and allopatric species of the *obscura* group of *Drosophila*. *Anim. Behav.*, **16**, 79–87.

44. — — Pavlovsky, O., and Spassky, B. 1964. The superspecies *Drosophila paulistorum*. *Proc. natn. Acad. Sci. U.S.A.*, **51**, 3–9.

45. — and Koller, P. Ch. 1938. An experimental study of sexual isolation in *Drosophila*. *Biol. Zbl.*, **58**, 589–607.

46. — and Pavlovsky, O. 1967. Experiments on the incipient species of the *Drosophila paulistorum* complex. *Genetics*, **55**, 141–156.

47. — — Spassky, B., and Spassky, N. 1955. Genetics of natural populations. XXIII. Biological role of deleterious recessives in populations of *Drosophila pseudoobscura*. *Genetics*, **40**, 781–796.

48. — and Wright, S. 1943. Genetics of natural populations. X. Dispersion rates in *Drosophila pseudoobscura*. *Genetics*, **28**, 304–340.

49. East, E. M. and Mangelsdorf, A. J. 1925. A new interpretation of the hereditary behavior of self-sterile plants. *Proc. natn. Acad. Sci. U.S.A.*, **11**, 166–171.

50. Ehrman, L. 1964. Genetic divergence in M. Vetukhiv's experimental populations of *Drosophila pseudoobscura*. I. Rudiments of sexual isolation. *Genet. Res.*, **5**, 150–157.

51. — 1965. Direct observation of sexual isolation between allopatric and between sympatric strains of the different *Drosophila paulistorum* races. *Evolution*, **19**, 459–464.

52. — 1969. The sensory basis of mate selection in *Drosophila*. *Evolution*, **23**, 59–64.

53. — and Petit, C. 1968. Genotype frequency and mating success in the *willistoni* species group of *Drosophila*. *Evolution*, **22**, 649–658.

54. — Spassky, B., Pavlovsky, O., and Dobzhansky, Th. 1965. Sexual selection, geotaxis, and chromosomal polymorphism in experimental populations of *Drosophila pseudoobscura*. *Evolution*, **19**, 337–346.

55. — and Spiess, E. B. 1969. Rare-type mating advantage in *Drosophila*. *Am. Nat.*, **103**, 675–680.

56. Elens, A. A. and Wattiaux, J. M. 1964. Direct observation of sexual isolation. *Drosoph. Inf. Serv.*, **39**, 118–119.

57. Emerson, S. 1939. A preliminary survey of the *Oenothera organensis* population. *Genetics*, **24**, 524–537.

58. — 1940. Growth of incompatible pollen tubes in *Oenothera organensis*. *Bot. Gaz.*, **101**, 890–911.

59. Ewens, W. J. 1965. Further notes on the evolution of dominance. *Heredity*, **20**, 443–450.
60. Fisher, R. A. 1928a. The possible modification of the response of the wild type to recurrent mutations. *Am. Nat.*, **62**, 115–126.
61. — 1928b. Two further notes on the origin of dominance. *Am. Nat.*, **62**, 571–574.
62. — 1930a. *The Genetical Theory of Natural Selection.* Clarendon Press, Oxford.
63. — 1930b. The evolution of dominance in certain polymorphic species. *Am. Nat.*, **64**, 385–406.
64. — 1935. Dominance in poultry. *Phil. Trans. R. Soc. B*, **225**, 197–226.
65. — 1938. Dominance in poultry. Feathered feet, rose comb, internal pigment and pile. *Proc. R. Soc. B*, **125**, 25–48.
66. — 1958. *The Genetical Theory of Natural Selection*, 2nd ed. Dover, New York.
67. — and Holt, S. B. 1944. The experimental modification of dominance in Danforth's short-tailed mutant mice. *Ann. Eugen.*, **12**, 102–120.
68. Ford, E. B. 1940a. Polymorphism and taxonomy. In: *The New Systematics*, ed. Huxley, J. Clarendon Press, Oxford.
69. — 1940b. Genetic research in the Lepidoptera. *Ann. Eugen.*, **10**, 227–252.
70. Gershenson, S. 1928. A new sex-ratio abnormality in *Drosophila obscura*. *Genetics*, **13**, 488–507.
71. Gibson, J. B. and Thoday, J. M. 1964. Effects of disruptive selection. IX. Low selection intensity. *Heredity*, **19**, 125–130.
72. Goldberg, A. L. and Wittes, R. E. 1966. Genetic code: aspects of organization. *Science*, **153**, 420–424.
73. Goodhart, C. 1963. 'Area effects' and non-adaptive variation between populations of *Cepaea* (Mollusca). *Heredity*, **18**, 459–465.
74. Grant, V. 1963. *The Origin of Adaptations.* Columbia University Press, New York.
75. — 1966. The selective origin of incompatibility barriers in the plant genus *Gilia*. *Am. Nat.*, **100**, 99–118.
76. Haldane, J. B. S. 1930. A note on Fisher's theory of the origin of dominance and on a correlation between dominance and linkage. *Am. Nat.*, **64**, 87–90.
77. — 1939. The theory of the evolution of dominance. *J. Genet.*, **37**, 365–374.
78. — 1949. Disease and evolution. *Ricerca scient.*, *Suppl.* **19**, 68–76.
79. — and Jayakar, S. D. 1963. Polymorphism due to selection depending on the composition of a population. *J. Genet.*, **58**, 318–323.
80. Hamilton, W. D. 1967. Extraordinary sex ratios. *Science*, **156**, 477–488.

81. Harding, J., Allard, R. W., and Smeltzer, D. G. 1966. Population studies in predominantly self-pollinated species. 9. Frequency-dependent selection in *Phaseolus lunatus*. *Proc. natn. Acad. Sci. U.S.A.*, **56**, 99–104.

82. Harris, H. 1966. Enzyme polymorphisms in man. *Proc. R. Soc. B*, **164**, 298–310.

83. — 1969. Enzyme and protein polymorphism in human populations. *Br. med. Bull.*, **25**, 5–13.

84. Henslee, E. D. 1966. Sexual isolation in a parthenogenetic strain of *Drosophila mercatorum*. *Am. Nat.*, **100**, 191–197.

85. Hickey, W. A. and Craig, G. B. 1966. Genetic distortion of sex ratio in a mosquito, *Aedes aegypti*. *Genetics*, **53**, 1177–1196.

86. Hoenigsberg, H. F. and Koref-Santibañez, S. K. 1960. Courtship and sensory preferences in inbred lines of *Drosophila melanogaster*. *Evolution*, **14**, 1–7.

87. Hubby, J. L. and Lewontin, R. C. 1966. A molecular approach to the study of genic heterozygosity in natural populations. I. The number of alleles at different loci in *Drosophila pseudoobscura*. *Genetics*, **54**, 577–594.

88. Hutchinson, J. B. 1946. The crinkled dwarf allelomorph series in the New World cottons. *J. Genet.*, **47**, 178–207.

89. Huxley, J. S. 1942. *Evolution: The Modern Synthesis*. Harper and Brothers, New York and London.

90. — 1955. Morphism and evolution. *Heredity*, **9**, 1–52.

91. Jain, S. K. and Bradshaw, A. D. 1966. Evolutionary divergence among adjacent plant populations. 1. The evidence and its theoretical analysis. *Heredity*, **21**, 407–441.

92. Kessler, S. 1966. Selection for and against ethological isolation between *Drosophila pseudoobscura* and *Drosophila persimilis*. *Evolution*, **20**, 634–645.

93. Kettlewell, H. B. D. 1956. A résumé of investigations on the evolution of melanism in the Lepidoptera. *Proc. R. Soc. B*, **145**, 297–303.

94. — 1958. A survey of the frequencies of *Biston betularia* (L.) (Lep.) and its melanic forms in Great Britain. *Heredity*, **12**, 51–72.

95. — 1961. The phenomenon of industrial melanism in Lepidoptera. *A. Rev. Ent.*, **6**, 245–262.

96. Kimura, M. 1968. Genetic variability maintained in a finite population due to mutational production of neutral and nearly neutral isoalleles. *Genet. Res.*, **11**, 247–269.

97. — and Crow, J. F. 1964. The number of alleles that can be maintained in a finite population. *Genetics*, **49**, 725–738.

98. King, J. C. 1947. Interspecific relationships within the *guarani* group of *Drosophila*. *Evolution*, **1**, 143–153.

99. King, J. L. 1967. Continuously distributed factors affecting fitness. *Genetics*, **55**, 483–492.

100. KING, J. L. and JUKES, T. H. 1969. Non-Darwinian evolution. *Science*, **164**, 788–798.

101. KNIGHT, G. R., ROBERTSON, A., and WADDINGTON, C. H. 1956. Selection for sexual isolation within a species. *Evolution*, **10**, 14–22.

102. KOJIMA, K. and TOBARI, Y. N. 1969. The pattern of viability changes associated with genotype frequency at the alcohol dehydrogenase locus in a population of *Drosophila melanogaster*. *Genetics*, **61**, 201–209.

103. — and YARBROUGH, K. M. 1967. Frequency-dependent selection at the Esterase 6 locus in *Drosophila melanogaster*. *Proc. natn. Acad. Sci. U.S.A.*, **57**, 645–649.

104. KOOPMAN, K. F. 1950. Natural selection for reproductive isolation between *Drosophila pseudoobscura* and *Drosophila persimilis*. *Evolution*, **4**, 135–148.

105. KOREF-SANTIBAÑEZ, S. K. and WADDINGTON, C. H. 1958. The origin of sexual isolation between different lines within a species. *Evolution*, **12**, 485–493.

106. LAMOTTE, M. 1959. Polymorphism of natural populations of *Cepaea nemoralis*. *Cold Spring Harb. Symp. quant. Biol.*, **24**, 65–86.

107. LEVENE, H. 1953. Genetic equilibrium when more than one ecological niche is available. *Am. Nat.*, **87**, 331–333.

108. — PAVLOVSKY, O., and DOBZHANSKY, TH. 1954. Interaction of the adaptive values in polymorphic experimental populations of *Drosophila pseudoobscura*. *Evolution*, **8**, 335–349.

109. LEVIN, D. A. and KERSTER, H. W. 1967. Natural selection for reproductive isolation in *Phlox*. *Evolution*, **21**, 679–687.

110. LEVINS, R. and MACARTHUR, R. 1966. The maintenance of genetic polymorphism in a spatially heterogeneous environment: variations on a theme by Howard Levene. *Am. Nat.*, **100**, 585–589.

111. LEWONTIN, R. C. 1955. The effects of population density and composition on viability in *Drosophila melanogaster*. *Evolution*, **9**, 27–41.

112. — 1958. A general method for investigating the equilibrium of gene frequency in a population. *Genetics*, **43**, 419–434.

113. — 1967. An estimate of average heterozygosity in man. *Am. J. hum. Genet.*, **19**, 681–685.

114. — and HUBBY, J. L. 1966. A molecular approach to the study of genic heterozygosity in natural populations. II. Amount of variation and degree of heterozygosity in natural populations of *Drosophila pseudoobscura*. *Genetics*, **54**, 595–609.

115 LI, C. C. 1955. *Population Genetics*. University of Chicago Press, Chicago.

116. LITTLEJOHN, M. J. 1965. Premating isolation in the *Hyla ewingi* complex (Anura: Hylidae). *Evolution*, **19**, 234–243.

117. LITTLEJOHN, M. J. and LOFTUS-HILLS, J. J. 1968. An experimental evaluation of premating isolation in the *Hyla ewingi* complex (Anura: Hylidae). *Evolution*, **22**, 659–663.

118. LORKOVIĆ, Z. 1958. Some peculiarities of spatially and sexually restricted gene exchange in the *Erebia tyndarus* group. *Cold Spring Harb. Symp. quant. Biol.*, **23**, 319–325.

119. MANWELL, C. and BAKER, C. M. A. 1968. Genetic variation of isocitrate, malate, and 6-phosphogluconate dehydrogenases in snails of the genus *Cepaea*—introgressive hybridization, polymorphism and pollution? *Comp. Biochem. Physiol.*, **26**, 195–209.

120. MATHER, K. 1955. Polymorphism as an outcome of disruptive selection. *Evolution*, **9**, 52–61.

121. MAYNARD SMITH, J. 1962. Disruptive selection, polymorphism and sympatric speciation. *Nature*, **195**, 60–62.

122. — 1966. Sympatric speciation. *Am. Nat.*, **100**, 637–650.

123. MAYO, O. 1966. On the evolution of dominance. *Heredity*, **21**, 499–511.

124. MAYR, E. 1931. Birds collected during the Whitney South Sea Expedition. XII. Notes on *Halcyon chloris* and some of its subspecies. *Am. Mus. Novit.*, **No. 469**, 1–10.

125. — 1963. *Animal Species and Evolution*. Harvard University Press, Cambridge, Mass.

126. MCNEILLY, T. and ANTONOVICS, J. 1968. Evolution in closely adjacent plant populations. IV. Barriers to gene flow. *Heredity*, **23**, 205–218.

127. MECHAM, J. S. 1960. Introgressive hybridization between two southeastern treefrogs. *Evolution*, **14**, 445–457.

128. MEISE, W. 1928. Die Verbreitung der Aaskrähe (Formenkreis *Corvus corone* L.). *J. Orn., Lpz.*, **76**, 1–203.

129. MERRELL, D. J. 1960. Mating preferences in *Drosophila*. *Evolution*, **14**, 525–526.

130. MICHAUD, T. C. 1964. Vocal variation in two species of chorus frogs, *Pseudacris nigrita* and *Pseudacris clarki*, in Texas. *Evolution*, **18**, 498–506.

131. MILKMAN, R. D. 1967. Heterosis as a major cause of heterozygosity in nature. *Genetics*, **55**, 493–495.

132. MOORE, J. A. 1957. An embryologist's view of the species concept. In: *The Species Problem*, ed. Mayr, E. *Publs Am. Ass. Advmt Sci.*, **No. 50**, 325–338.

133. MORGAN, T. H. 1929. Variability of eyeless. *Publs Carnegie Instn*, **No. 399**, 139–168.

134. MOURANT, A. E. 1961. The significance of the red cell antigens. In: *Functions of the Blood*, eds. MacFarlane, R. G. and Robb-Smith, A. H. T. Academic Press, New York.

135. MULCAHY, D. L. 1967. Optimal sex ratio in *Silene alba*. *Heredity*, **22**, 411–423.

136. MULLER, H. J. 1942. Isolating mechanisms, evolution and temperature. *Biol. Symp.*, **6**, 71–125.

137. MULLER, H. J. 1950. Our load of mutations. *Am. J. hum. Genet.*, **2**, 111–176.
138. MURRAY, J. and CLARKE, B. 1966. The inheritance of polymorphic shell characters in *Partula* (Gastropoda). *Genetics*, **54**, 1261–1277.
139. —— 1968. Partial reproductive isolation in the genus *Partula* (Gastropoda) on Moorea. *Evolution*, **22**, 684–698.
140. O'BRIEN, S. J. and MACINTYRE, R. J. 1969. An analysis of gene-enzyme variability in natural populations of *Drosophila melanogaster* and *D. simulans*. *Am. Nat.*, **103**, 97–113.
141. O'DONALD, P. 1967a. On the evolution of dominance, over-dominance and balanced polymorphism. *Proc. R. Soc. B*, **168**, 216–228.
142. — 1967b. The evolution of selective advantage in a deleterious mutation. *Genetics*, **56**, 399–404.
143. — and PILECKI, C. 1970. Polymorphic mimicry and natural selection. *Evolution*, **24**, 395–401.
144. PARSONS, P. A. and BODMER, W. F. 1961. The evolution of over-dominance: natural selection and heterozygote advantage. *Nature*, **190**, 7–12.
145. PATERNIANI, E. 1969. Selection for reproductive isolation between two populations of maize, *Zea mays* L. *Evolution*, **23**, 534–547.
146. PATTERSON, J. T. and STONE, W. S. 1952. *Evolution in the Genus Drosophila*. Macmillan, New York.
147. PETIT, C. 1951. Le rôle de l'isolement sexuel dans l'évolution des populations de *Drosophila melanogaster*. *Bull. biol. Fr. Belg.*, **85**, 392–418.
148. — 1954. L'isolement sexuel chez *Drosophila melanogaster*. Étude du mutant *white* et de son allélomorphe sauvage. *Bull. biol. Fr. Belg.*, **88**, 435–443.
149. — 1968. Le rôle des valeurs sélectives variables dans le maintien du polymorphisme. *Bull. Soc. zool. Fr.*, **93**, 187–208.
150. PIMENTEL, D., SMITH, G. J. C., and SOANS, J. 1967. A population model of sympatric speciation. *Am. Nat.*, **101**, 493–504.
151. PLATT, A. P. and BROWER, L. P. 1968. Mimetic versus disruptive coloration in intergrading populations of *Limenitis arthemis* and *astyanax* butterflies. *Evolution*, **22**, 699–718.
152. POPHAM, E. J. 1941. The variation in the colour of certain species of *Arctocorisa* (Hemiptera, Corixidae) and its significance. *Proc. zool. Soc. Lond. A*, **111**, 135–172.
153. — 1942. Further experimental studies on the selective action of predators. *Proc. zool. Soc. Lond. A*, **112**, 105–117.
154. POULTON, E. B. 1909. Mimicry in the butterflies of North America. *Ann. ent. Soc. Am.*, **2**, 203–242.
155. PRAKASH, S. and LEWONTIN, R. C. 1968. A molecular approach to the study of genic heterozygosity in natural populations. III. Direct evidence of coadaptation in gene arrangements of *Drosophila*. *Proc. natn. Acad. Sci. U.S.A.*, **59**, 398–405.

156. PRAKASH, S., LEWONTIN, R. C., and HUBBY, J. L. 1969. A molecular approach to the study of genic heterozygosity in natural populations. IV. Patterns of genic variation in central, marginal and isolated populations of *Drosophila pseudoobscura*. *Genetics*, **61**, 841–858.

157. REIGHARD, J. 1908. An experimental field-study of warning coloration in coral-reef fishes. *Publs Carnegie Instn*, **No. 103**, 257–325.

158. REMINGTON, C. L. 1958. Genetics of populations of Lepidoptera. *Proc. 10th Int. Congr. Ent.*, **2**, 787–805.

159. — 1968. Suture-zones of hybrid interaction between recently joined biotas. *Evol. Biol.*, **2**, 321–428.

160. RICK, C. M. 1963. Barriers to interbreeding in *Lycopersicon peruvianum*. *Evolution*, **17**, 216–232.

161. ROBERTSON, F. W. 1966a. The ecological genetics of growth in *Drosophila*. *Genet. Res.*, **8**, 165–179.

162. — 1966b. A test of sexual isolation in *Drosophila*. *Genet. Res.*, **8**, 181–187.

163. SCHAD, G. A. 1966. Immunity, competition and natural regulation of helminth populations. *Am. Nat.*, **100**, 359–364.

164. SCHARLOO, W., DEN BOER, M., and HOOGMOED, M. S. 1967. Disruptive selection on sternopleural chaeta number in *Drosophila melanogaster*. *Genet. Res.*, **9**, 115–118.

165. SELANDER, R. K., HUNT, W. G., and YANG, S. Y. 1969. Protein polymorphism and genic heterozygosity in two European subspecies of the house mouse. *Evolution*, **23**, 379–390.

166. — YANG, S. Y., LEWONTIN, R. C., and JOHNSON, W. E. 1970. Genetic variation in the horseshoe crab (*Limulus polyphemus*), a phylogenetic 'relic'. *Evolution*, **24**, 402–414.

167. SHAW, R. F. 1961. The effect of polygamy and infanticide on the sex ratio. *Am. J. phys. Anthrop.*, **19**, 79–84.

168. SHEPPARD, P. M. 1958. *Natural Selection and Heredity*. Hutchinson, London.

169. — 1959. The evolution of mimicry: a problem in ecology and genetics. *Cold Spring Harb. Symp. quant. Biol.*, **24**, 131–140.

170. — and COOK, L. M. 1962. The manifold effects of the *medionigra* gene of the moth *Panaxia dominula* and the maintenance of a polymorphism. *Heredity*, **17**, 415–426.

171. — and FORD, E. B. 1966. Natural selection and the evolution of dominance. *Heredity*, **21**, 139–147.

172. SIBLEY, C. G. 1957. The evolutionary and taxonomic significance of sexual dimorphism and hybridization in birds. *Condor*, **59**, 166–191.

173. — 1961. Hybridization and isolating mechanisms. In: *Vertebrate Speciation*, ed. Blair, W. F. University of Texas Press, Austin.

174. SIMONSEN, M. and HARRIS, R. J. C. 1956. Induced susceptibility of turkeys to the Rous virus after treatment in the embryonic stage

with normal chicken blood. *Acta path. microbiol. scand.*, **Suppl. 111**, 53–54.

175. SMITH, H. M. 1965. More evolutionary terms. *Syst. Zool.*, **14**, 57–58.

176. SOKAL, R. R. and HUBER, I. 1963. Competition among genotypes in *Tribolium castaneum* at varying densities and gene frequencies (the *sooty* locus). *Am. Nat.*, **97**, 169–184.

177. — and KARTEN, I. 1964. Competition among genotypes in *Tribolium castaneum* at varying densities and gene frequencies (the *black* locus). *Genetics*, **49**, 195–211.

178. SPIESS, E. B. 1957. Relation between frequencies and adaptive values of chromosomal arrangements in *Drosophila persimilis*. *Evolution*, **11**, 84–93.

179. SPIESS, L. D. and SPIESS, E. B. 1969. Minority advantage in interpopulational matings of *Drosophila persimilis*. *Am. Nat.*, **103**, 155–172.

180. STEPHENS, S. G. 1946. The genetics of 'corky'. I. The new world alleles and their possible role as an interspecific isolating mechanism. *J. Genet.*, **47**, 150–161.

181. SUBAK-SHARPE, H., SHEPHERD, W. M., and HAY, J. 1966. Studies on sRNA coded by herpes virus. *Cold Spring Harb. Symp. quant. Biol.*, **31**, 583–594.

182. SVED, J. A., REED, T. E., and BODMER, W. F. 1967. The number of balanced polymorphisms that can be maintained in a natural population. *Genetics*, **55**, 469–481.

183. TAYLOR, J. W. 1914. *Monograph of the Land and Freshwater Mollusca of the British Isles*. Taylor Brothers, Leeds.

184. THODAY, J. M. 1959. Effects of disruptive selection. I. Genetic flexibility. *Heredity*, **13**, 187–203.

185. — and GIBSON, J. B. 1962. Isolation by disruptive selection. *Nature*, **193**, 1164–1166.

186. — — 1970. The probability of isolation by disruptive selection. *Am. Nat.*, **104**, 219–230.

187. TINBERGEN, L. 1960. The natural control of insects in pinewoods. I. Factors influencing the intensity of predation by song-birds. *Archs neerl. Zool.*, **13**, 265–336.

188. TOBARI, Y. N. and KOJIMA, K. 1967. Selective modes associated with inversion karyotypes in *Drosophila ananassae*. I. Frequency-dependent selection. *Genetics*, **57**, 179–188.

189. UZZELL, T. M. 1964. Relations of the diploid and triploid species of the *Ambystoma jeffersonianum* complex (Amphibia, Caudata). *Copeia*, **1964**, 257–300.

190. VAURIE, C. 1957. Systematic notes on palearctic birds. No. 26. Paridae: the *Parus caeruleus* complex. *Am. Mus. Novit.*, **No. 1833**, 1–15.

191. WALLACE, A. R. 1889. *Darwinism*. Macmillan, London.

192. Wallace, B. 1954. Genetic divergence of isolated populations of *Drosophila melanogaster*. *Proc. IX Int. Congr. Genet., Caryologia*, **6** (Suppl.), 761–764.
193. — 1958. The average effect of radiation-induced mutations on viability in *Drosophila melanogaster*. *Evolution*, **12**, 532–556.
194. — 1963. Further data on the overdominance of induced mutations. *Genetics*, **48**, 633–651.
195. — 1966. *Chromosomes, Giant Molecules, and Evolution*. W. W. Norton, New York.
196. — 1968. *Topics in Population Genetics*. W. W. Norton, New York.
197. Watson, G. F. and Martin, A. A. 1968. Postmating isolation in the *Hyla ewingi* complex (Anura: Hylidae). *Evolution*, **22**, 664–666.
198. West, D. 1964. Polymorphism in the isopod *Sphaeroma rugicauda*. *Evolution*, **18**, 671–684.
199. Williams, G. C. 1966. *Adaptation and Natural Selection*. Princeton University Press, Princeton, N.J.
200. Williamson, M. H. 1960. On the polymorphism of the moth *Panaxia dominula* (L.). *Heredity*, **15**, 139–151.
201. Wright, S. 1925. The factors of the albino series of guinea-pigs and their effects on black and yellow pigmentation. *Genetics*, **10**, 223–260.
202. — 1929a. Fisher's theory of dominance. *Am. Nat.*, **63**, 274–279.
203. — 1929b. The evolution of dominance: comment on Dr. Fisher's reply. *Am. Nat.*, **63**, 556–561.
204. — 1931. Evolution in mendelian populations. *Genetics*, **16**, 97–159.
205. — 1939. The distribution of self-sterility alleles in populations. *Genetics*, **24**, 538–552.
206. — 1964. The distribution of self-incompatibility alleles in populations. *Evolution*, **18**, 609–619.
207. Yarbrough, K. M. and Kojima, K. 1967. The mode of selection at the polymorphic Esterase 6 locus in cage populations of *Drosophila melanogaster*. *Genetics*, **57**, 677–686.

References added in proof

208. Berger, E. M. 1970. A comparison of gene-enzyme variation between *Drosophila melanogaster* and *D. simulans*. *Genetics*, **66**, 677–683.
209. Clarke, B. 1969. The evidence for apostatic selection. *Heredity*, **24**, 347–352.
210. Crosby, J. L. 1970. The evolution of genetic discontinuity: computer models of the selection of barriers to interbreeding between subspecies. *Heredity*, **25**, 253–297.
211. Ohh, B. K. and Sheldon, B. L. 1970. Selection for dominance of Hairy-wing (*Hw*) in *Drosophila melanogaster* I. Dominance at different levels of phenotype. *Genetics*, **66**, 517–540.
212. Robertson, A. 1970. A note on disruptive selection experiments in *Drosophila*. *Am. Nat.*, **104**, 561–569.

Index